Relax  to the Max

Relax to the Max

60 CANDLES, SCENTS, SOAPS & POTPOURRI CRAFTS
TO CREATE YOUR OWN BLISS

BY ROSEVITA WARDA, M. LOU LUCHSINGER,
MARIE BROWNING & DAWN CUSICK

Sterling Publishing Co., Inc.
New York

Every effort has been made to ensure that the information presented is accurate. Since we have no control over physical conditions, individual skills, or chosen tools and products, the publisher disclaims any liability for injuries, losses, untoward results, or any other damages which may result from the use of the information in this book. Thoroughly read the instructions for all products used to complete the projects in this book, paying particular attention to all cautions and warnings shown for that product to ensure their proper and safe use.

Library of Congress Cataloging-in-Publication Data Available

2   4   6   8   10   9   7   5   3   1

Published by Sterling Publishing Co., Inc.
387 Park Avenue South, New York, NY 10016
© 2005 by Sterling Publishing Co., Inc
This book is comprised of material from the following Sterling Publishing Co., Inc. titles:
Aromatic Candles © 2001 by Rosevita Warda
Indulge Yourself with Aromatherapy © 2000 by Prolific Impressions, Inc.
300 Handcrafted Soaps © 2002 by Prolific Impressions, Inc.
Potpourri Crafts © 1992 by Altamont Press

Cover and interior design by 3+Co. (www.threeandco.com)
Photography on pages 6, 14, 59, 63-69, 71-75, 83, 117, 118, 120 © Getty Images
Photography on pages 15-17, 21, 60, 62, 70, 116, 119 © Corbis

Distributed in Canada by Sterling Publishing
c/o Canadian Manda Group, 165 Dufferin Street
Toronto, Ontario, Canada M6K 3H6
Distributed in Great Britain by Chrysalis Books Group PLC
The Chrysalis Building, Bramley Road, London W10 6SP, England
Distributed in Australia by Capricorn Link (Australia) Pty. Ltd.
P.O. Box 704, Windsor, NSW 2756, Australia

Printed in China
All rights reserved

Sterling ISBN 1-4027-1931-0

TABLE OF CONTENTS

$\mathcal{S}$top and smell the roses. A lovely idea, but really, who has the time? As a modern woman, you are constantly on the go, balancing your time between work, family, friends, and other areas of your life. But you need to remember to make time for yourself—finding time to release the cares of the day renews your body and spirit. And *Relax to the Max* is all about helping you do just that through aromatherapy. By making and using the fragrant candles, scents, soaps, and potpourris in this book, you will be able to incorporate the rejuvenating powers of aromatherapy into your daily routines from morning to night. These delightful scented creations will inspire you to find peace, to enjoy yourself, and to make the most of each day. The power of scent is tremendous.

Aromatherapy is a modern term for a healing art that is ages old. It is a way to improve the quality of life on a physical, emotional, and spiritual level. The basis of aromatherapy is the use of essential oils, the vital life essence of aromatic plants and flowers distilled in a highly concentrated form.

The Chinese first used essential oils for medicinal purposes as early as 4000 B.C. The Egyptians incorporated aromatics in their rituals and healing, including massage cosmetics and embalming. Many of their formulas were carved into the stone walls of Egyptian temples. However, the first guide to describe the numerous uses for plant oils was not published until the seventeenth century. In the nineteenth century chemical substitutes for essential oils became popular and almost halted the use of pure, natural oils.

The rediscovery of the healing properties of essential oils is, for the most part, thanks to Rene-Maurice Gattefossé, a French chemist working in his family's perfume business, who, in 1928, coined the term "aromatherapy." At first his research was confined to the cosmetic uses of essential oils, but in 1910 he discovered by accident that lavender was able to heal a severe burn rapidly and helped prevent scarring. His discovery led to the use of lavender oil to treat burn victims during World War I. Fascinated with the therapeutic possibilities of essential oils, Gattefossé continued his research on their healing properties for the rest of his life.

Gattefossé's work generated a great deal of interest in the healing properties of essential oils, particularly in France and Italy. Not only were fragrance oils found to heal burns, wounds, and other skin conditions and to help strengthen the immune system, they were also found to relieve mental conditions such as anxiety and depression.

During the last thirty years, increasing attention has been paid to aromatherapy's contribution to health and well-being. A number of hotels, airports, and resort areas now offer aromatherapy massage and products to their clients. And aromatherapy has taken the cosmetics industry by storm. Many cosmetic companies have introduced a line of aromatherapy products that promotes the beneficial aspects of fragrances. Although no medical claims can be made, it is certainly no surprise that a relaxing, scented bath can soothe and refresh.

Breathing in the fragrant aromas of essential oils elicits physical, emotional, and spiritual responses that promote a sense of health and well-being. It is also believed that the molecules of an essential oil permeate the skin and are carried by the lymphatic and circulatory systems to the inner organs to provide the same beneficial effects. Whether the oil is absorbed through the skin or inhaled, once it is in the bloodstream and body fluids, it works therapeutically, however small the dose. The chemicals in essential oils unlock the body's ability to heal. They influence all systems of the body, from tissues and organs, to body fluids and cells, as well as emotional and spiritual states.

The results of aromatherapy are very individual; no two people are affected by the same essential oil in exactly the same way.

Even the same person can be affected differently by the same oil depending on the surroundings, time of day, or mood.

The key to using aromatherapy to improve our quality of life is to find the scents, unique for each individual, that evoke positive sensory feelings and emotions, and then to introduce those scents into our everyday life to enhance our well-being. Natural scents keep us connected to the earth, grounding us, sparking memories and emotions, and healing the spirit.

Scent Secrets

To be a bona fide goddess of scented crafts, you first need to know the basics about working with fragrances. Fragrance oils are the core of relaxing candles, bath oils, soaps, and potpourri. Although some scents can come from additives, such as rose petals in potpourris or oatmeal in soaps, they will not be strong enough for a truly blissful aromatic creation. For the best results, the two types of oils to use are essential oils, the naturally scented oils of fresh flowers and herbs produced by plants, and synthetic fragrance oils, which are artificially produced scents. This chapter lets you in on all of the secrets about oils, scents, and blends, as well as the best ways to buy and store your oils and safety precautions.

Essential Oils

Of the many thousands of plants in the world, only two hundred produce the aromatic essential oils used in the art of perfumery. Essential oils are the pure essences of healing and aromatic plants. They are the very soul of the plant, a precious extraction. The oil is contained in highly specialized glands in the foliage, flower, or other plant material. In fact, roots, stalks, bark, leaves, flowers, blossoms, seeds, nuts, fruits, and resins have all been used to obtain the plant oil. The plants that contain essential oils are found mainly in hot, dry habitats. At certain times of the day, and particular times of the year, the essential oils are present at optimum levels, which is the best time for harvest and distillation.

Essential oils are highly concentrated and must be diluted before they can safely be used in craft projects or applied to the skin (1% or less is considered to be a safe level). Too much of an essential oil can cause severe skin irritations. People with sensitive skin or allergies should be careful when using essential oils. Pure essences are more expensive than synthetics, since it often requires hundreds of pounds of plant matter to extract even a tiny amount of essential oil. Additionally, you must be more careful when storing essential oils than synthetically produced fragrances because they are more potent.

The action of an essential oil on the body is holistic, combining both physical and mental aspects.

Essential oils work together with all aspects of the body, strengthening rather than weakening it so that the entire body may aid in the healing and restorative process. Essential oils are extremely flexible in their applications. Their ability to affect people on physical, emotional, and psychological levels is a special element unmatched by other healing arts.

SYNTHETIC FRAGRANCE OILS

Many fragrance oils are synthetic and are not derived from specific plants. These synthetic fragrance oils are cheaper to produce than essential oils and are, therefore, less expensive. They also come in a wider range of scents and blends than essential oils. Many high-quality fragrance oils are actually blends of essential oils.

Be aware that some synthetic fragrance oils contain ingredients that are suspected to be carcinogenic or responsible for various allergic reactions, and are toxic or irritating to the skin. Also, keep in mind while they have a pleasant aroma, synthetic fragrance oils lack the healing properties of all-natural essential oils.

Both essential and synthetic fragrance oils can be used in the recipes in this book. In this book both essential and synthetic fragrance oils are collectively referred to as fragrance oils.

BUYING FRAGRANCE OILS

Fragrance oils are the most expensive ingredients you will buy for your scented creation, but they are the most important. Let your budget and your nose dictate what oils to purchase. Do not buy cheap oils or extracts; you will be disappointed in the finished product if you do. Your nose can help you determine the quality of a fragrance oil. Oils diluted or cut with alcohol tend to smell alike and will have a sharp bite to them. Fragrance oils that contain alcohol adversely affect your finished product as the alcohol pushes out moisture.

Quality, uncut fragrance oils have a smooth, yet concentrated aroma. When making scented creations, you will use more of a lesser quality fragrance oil to achieve the same results produced by a high-quality fragrance oil. Since the purity of the fragrance is often what invokes a positive response to a finished product, good-quality fragrance oils are a wise investment.

Fragrance oils come in various grades and varieties, so you will need to literally sniff them out. Lavender oil, for example, can vary greatly in quality and scent, depending on plant origin, distiller's expertise, and the quality of storage. When buying fragrance oils, carry a bag of freshly ground coffee and sniff it occasionally to help clear your nose. To make sure you buy the best fragrance oils, purchase oils from companies that have an established reputation in the aromatherapy field. Also, read the fragrance oil label. Does it ensure purity? Does it state the botanical name, country of origin, form of extraction, and how the plant was cultivated? Be aware that scents such as rain, lily of the valley, and raspberry are synthetic and cannot be captured naturally.

Since essential oils are overly strong in their pure form, they may be sold in dilution with a natural base oil, which is fine provided they are labeled as such. If you have any questions as to whether or not the essential oil you purchase is at full strength, place a drop of oil on a piece of paper and allow it to evaporate completely. The remaining stain should not look like an oily mark, which indicates that the essential oil was stretched with a vegetable oil.

Mixing your own scents to create lasting fragrant blends is not difficult. To blend, all you need to know is what scents you like and enjoy and which will work for you. Many recipes in this book use combinations of fragrance oils to create distinctive, enduring scents.

Treat yourself to an assorted selection of oils. The more oils you collect, the more experimenting and creating you can do. Smell the oils, either by passing the bottle directly below your nose, or by putting a few drops on a piece of blotting paper or on unscented tissue. Then sit back and observe the feelings and images the single oil generates. Do you perceive them as calming, energizing, grounding, or sensual? Become acquainted with the individual oils by using them until you become familiar with the various dimensions and qualities they possess.

Scents in the perfume trade are categorized into groups, such as citrus, spicy, herbal, fruity, floral, and earthy. Knowing these groups helps you decide how to mix your blends; for example, you may want a rose scent (from the floral category) with a slight undertone of clove (from the spice category). You will naturally choose scents you find most appealing. While some people enjoy rich, earthy, exotic scents, others love the refreshing, clean scents of citrus or herbal categories.

BASIC FRAGRANCE BLENDING

Fragrance blending, a centuries-old skill that creates bottles of luxury and dreams, is the most creative part of scent-crafting. When compatible fragrance oils are combined in a blend, a synergy is created, resulting in a combination that is more powerful than the sum of its parts.

Or they can help you focus and clear your head, such as frankincense, peppermint, grapefruit, cinnamon, chamomile, lavender, orange, and ylang-ylang. Fragrances can work as antiseptics, for example tea tree, eucalyptus, peppermint, and lavender.

A PARTIAL LIST OF FRAGRANCES AND THEIR SCENT GROUPS

CITRUS
Bergamot
Grapefruit
Lemon
Lemongrass
Lime
Mandarin
Petitgrain
Pink grapefruit
Sweet orange
Tangerine

SPICY
Cardamom
Cinnamon
Clove
Ginger
Nutmeg
Vanilla
Mixed spice

HERBAL
Basil
Bayberry
Chamomile
Cucumber

Eucalyptus
Juniper
Marjoram
Peppermint
Pine
Rosemary
Sage
Tea tree

FRUITY
Apricot
Blackberry
Blueberry
Coconut
Green apple
Kiwi
Mango
Melon
Mulberry
Pear

FLORAL
Jasmine
Lavender
Lilac
Lily of the valley

Neroli
Plumaria
Rose
Violet
Ylang-ylang

EARTHY
Amber
Frankincense
Honey
Musk
Patchouli
Sandalwood

BLENDED
Fragrances
Baby powder
Brown sugar
Buttery maple
Candy cane
Chocolate
Honey almond
Ocean
Rain
Sunflower

QUALITIES ATTRIBUTED TO ESSENTIAL OILS

Many essential oils have specific effects and qualities attributed to them. Researchers are divided on whether the benefits of aromatherapy are conferred by the aromas or by other properties of the oils. Lavender, sandalwood, honeysuckle, chamomile, ylang-ylang, tangerine, rose, and lemon verbena are thought to be peaceful and relaxing. While rosemary, peppermint, lemon, lime, jasmine, and honey are energizing. Fragrances can be stimulating and uplifting, such as bergamot, orange, jasmine, rosemary, lemon verbena, mints, sage, and pine.

Single-scent perfumes that smell like natural botanicals, such as a freshly sliced orange or a fresh sprig of peppermint, are very popular. Although these perfumes may smell like a single fruit, flower, or herb, they contain other scents that make them lasting and more charming. These other scents, called "notes," are divided into three main elements: main scents, blenders, and contrasting scents.

The main scent is the key, or dominant, scent in your blend that creates the overall aroma. These top notes are the first aromas your nose detects, but will dissipate quickly. Blenders are additional scents that enhance the main scent and form the middle notes, or body, of a blend. Some blends include scents from other categories. Successful blenders include lime, peppermint, lavender, rose, jasmine, sandalwood, vanilla, cinnamon, and honey. Contrasting scents liven up the blend and provide the low base notes that are the lingering scents. Take, for example, the tropical scent of coconut. The blender scent could be the floral notes of ylang-ylang, which enhances and sweetens the coconut aroma. The contrasting scent could be vanilla, which livens up the coconut without overpowering it, and ylang-ylang scents, and provides the lasting note.

A fixative is necessary to help give the blend a long-lasting quality and release the fragrance moderately. The fixative takes the place of the plants' cells and holds the scent. It can be unscented or add its own aroma to the blend. For example, in soap crafting, the fixative can be the soap base or dried botanical additives. In potpourri, the fixative can be powdered orrisroot, sandalwood bark, or patchouli leaves.

Test your fragrance blend by placing a few drops of each oil that comprise the blend on a paper towel. Let the oils mingle for a few hours, then evaluate the blend by sniffing. The scent of the oil blend will alter as it matures. If you are not certain about a blend, allow it to age for a day or two and smell it again.

The recipes call for a moderate amount of fragrance oils for a pleasing, modest aroma. You may wish to add a little more for a stronger, more powerful scent. Remember some oils are much stronger than others. Whatever the scent, make it stronger than you think, as some of it will dissipate.

TO BLEND FRAGRANCE OILS:

1. Select the fragrance oils you would like to blend. There are no rules about how many oils should be in a blend. Beginners should start with three or four oils.

2. Place an equal number of drops of each oil in a clean glass bottle. Be sure to use a separate dropper for each fragrance oil.

3. Shake the bottle and test your results.

4. Introduce additional drops of oil a few at a time until you achieve the desired effect.

5. Write down the fragrance oils you use and the number of drops of each oil. Also note your perception of the blend, when freshly blended and after maturing.

TIP: Be sure to use the correct cap and dropper for each oil so you do not spoil the oils by mixing up the lids.

STORING OILS

Store essential oils and blends in dark glass bottles and keep them in a cool, dark place.

Always label oils and blends clearly. When labeling blends, be specific. If there is insufficient room on the bottle for labeling, number the blend and record a corresponding number in your blend notebook.

Avoid storing your oils with rubber eye-dropper tops in the bottles, as the oils may turn the rubber to gum, which will ruin the essential oils.

Always pay attention to any safety information listed on the essential oil label. Some essential oils should not be used during pregnancy or with other medical conditions. This applies primarily to topical usage or if ingested, but it is something that you should be aware of.

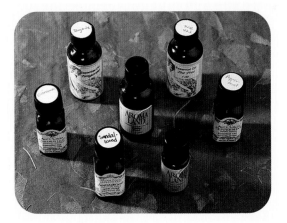

SAFETY TIPS

Essential oils are highly concentrated and potent substances, and it is important to understand a few cautionary guidelines when crafting with them.

- Do not take essential oils internally.
- Always work in an uncluttered workspace with good ventilation. Avoid hot rooms or areas with direct sunlight, as fragrance oils can evaporate quickly.
- If a fragrance oil is called a potpourri oil, candle scent, or flavored extract, do not use it in soap or bath products because they are not safe to use on your skin.
- Avoid all essential oils, natural herbal products, and bath salts during pregnancy.
- Fragrance oils should always be used diluted in a base; they are not perfumes that can be applied directly to your skin.
- Keep fragrance oils out of reach of children.
- Do not allow fragrance oils to come into contact with plastic. Certain oils will dissolve some plastics.
- Keep fragrance oils away from varnished or painted surfaces. For example, cinnamon oil can cleanly strip paint from furniture.

Candle Diva

Whoever said that candlemaking is for the pioneers never understood the power of a fragrant candle. But as a candle diva, you do. Or at least, you will. Candles ban stress and anxiety from your life; they energize you; and they increase your concentration. Whether you are worn out from work or need help waking up, turn to aromatherapy candles for a pick-me-up. By relaxing with an aromatherapy candle, you will improve your health as you use scents to soothe the demands and worries in your life. The citrusy aroma of a tangy bergamot, the sun-drenched essence of French lavender, or the sweet flavor of vanilla are guaranteed to make you feel as though you own the world. Aromatherapy is about following your nose into a mini-vacation—a moment of joy and harmony. The candles featured in this chapter are strong stimulants and will leave you rejuvenated, energized, and prepared to strut your stuff.

Creating Candles

Scenting candles with fragrance oils requires a little more effort and expense than ordinary candlemaking. Fragrance oils show different characteristics when added to wax and can change from batch to batch. Additionally, there is no guarantee that the scent of the blend you are using will be the same when the finished candle is burned. Candles made with fragrance oils may sputter, burn unevenly, or occasionally flare up. Pour a few votive-sized test candles and burn them down before using a blend in an elaborate and time-intensive candle project. Always keep in mind that a candle should never be left burning unattended; this is particularly important with an aromatherapy candle. The higher the concentration of oils used, the more important it is for one to test and carefully observe the candle as it burns.

WHAT YOU NEED

A few supplies and materials are needed for candlemaking. Most can be easily purchased at a craft store or found around the house. The essential oils may be purchased at health food or aromatherapy stores, or ordered by mail or through the Internet. Mail order or the Internet generally will be the most economical way to purchase essential oils, although the stores may be more convenient.

SUPPLIES

To create candles,
you will need the following
tools and supplies:

- CANDLE DYE
- CANDLE MOLDS
- CANDLE (OR CANDY) THERMOMETER
- CANDLE WAX
- CANDLE WICKS
- CANDLE WICK TABS
- DIPPING VAT
- DOUBLE BOILER
- FRAGRANCE OILS
- KITCHEN SCALE
- MEASURING SPOONS
- METAL POURING POT
- METAL SKEWER
- METAL SPOON
- MOLD RELEASE
- MOLD SEALER
- TAPE MEASURE
- WAX ADDITIVES

CANDLE WAXES

If you are making aromatic candles, keep in mind that the softer the wax, the more fragrant the candle seems. Soft wax does not seal its surface as tightly as hard wax, so it allows the scent to permeate the atmosphere around it. However, soft wax is more appropriate for container candles—tapers or free-standing candles would lose their shape and melt into a pool of wax.

Most manufacturers of candlemaking supplies offer wax compounds that are designed specifically for containers, molds, and tapers. This takes some of the guesswork out of trying to determine which additives are necessary for the style of candle that you are making.

Wax is most often sold by the block or sheet. To determine the amount of wax needed to fill the mold, fill the mold with water and measure the water. Three ounces of wax is equivalent to three-and-a-half ounces of water. To cut wax from the block or sheet, place the wax into a large garbage bag on a solid surface. Using a hammer and screwdriver, break wax into small chunks.

Natural Waxes

Beeswax is highly regarded among candlemakers because it burns beautifully with its own nurturing aroma. Beeswax is stickier than other waxes and requires a slightly larger wick. This may also cause problems when releasing a candle containing a high percentage of beeswax from the mold.

Beeswax works best when used in a container, dipped tapers, and rolled candles. Beeswax has a higher melting point (146°F) than most other waxes. It works well when mixed with lower melting-point waxes and used in container candles. Pure beeswax candles will burn directly down the center of the container, leaving much unburned wax on the container walls. Beeswax is available in its natural color or bleached white and is the most expensive of the waxes. Colored beeswax is also available, but may be a little more difficult to find. If your sheets of beeswax form a dusty-looking film on the surface, remove the dust by slightly warming the wax surface with a hair dryer.

Vegetable-based waxes are available and are made from different bases. Some are formulated from soybeans, jojoba beans, palm wax, and other vegetable bases. The majority of these waxes are made for container candles rather than pillars or tapers. They burn clean and are long lasting. These waxes may be difficult to find at the local craft store, but they are also available through mail order and the Internet.

Artificial Waxes

Paraffin is made from mineral oil and there is disagreement among candlemakers as to whether or not it is a natural wax. Those who believe that it is natural, argue that it is an organic substance. Those on the opposing side say it is artificial and inferior because it is made from petroleum that has been put through an elaborate refining process.

Paraffin is inexpensive and readily available. It is harder and more brittle than natural waxes and is a good choice for molded candles. It has a lower oil content and thus a higher melting point. It is also translucent and a favored choice for over dipping candles. However, paraffin is more likely to smoke and not burn as cleanly as other waxes. Some candlemakers have reported the fumes produced when melting paraffin can be nauseating. Also, the petroleum and carbon combustion ends up in the air and leaves black soot on walls and surfaces.

Gel wax is convenient to use because of its melt-and-pour capabilities as well as its visual appeal. It is made from mineral oil that is combined with substantial amounts of thermoplastic resin and butylated hydroxy toluene. Gel wax is not the most appropriate choice of wax for an aromatherapy candle and is somewhat incompatible with the idea of using natural essential oils in an artificial wax.

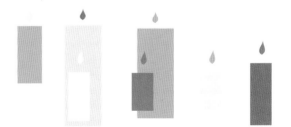

CANDLE MOLDS

Candle molds come in an endless variety of shapes and sizes and are made of acrylic, metal, plastic, or rubber. Molds are relatively inexpensive, and they can be used repeatedly.

Common household items such as cartons and containers also make excellent molds. Anything from paper milk cartons to smooth-sided aluminum cans can be used as molds. Simply make a hole in the bottom center of cartons and containers with an awl or drill to thread the wick through. Secure the wick as you would for a purchased mold.

To make your own mold from a piece of corrugated cardboard:

1. Coat the corrugated side of the cardboard with vegetable oil.

2. Shape the cardboard into a round, square, or triangular mold as desired with corrugated side in.

3. Wrap shipping tape around the outside of the cardboard to hold its shape.

4. Make a hole in the lid.

5. Using mold sealer, secure bottom edge of cardboard to a plastic container lid.

6. Thread wick through hole in the plastic lid and secure.

7. Pour melted wax into mold.

8. When wax is set, tear away cardboard to reveal candle.

Cleaning Molds

Candle molds should be clean and free from previous candle wax before using. Avoid scraping or scratching inside of molds when cleaning them, as scratches will mar future candles. Glass, plastic, or metal candle molds may be cleaned in any one of the following ways:

1. Fill sink with hot water and add liquid dish detergent. Allow candle molds to soak in water for ten minutes. Wash candle molds, rinse, and dry thoroughly.

2. Preheat oven to 200°F, and place paper towels on a cookie sheet. Place the candle molds upside down on the paper towels and place in the oven for seven to eight minutes. Remove the cookie sheet from the oven and wipe all excess wax from the inside of the candle molds.

3. Clean the metal molds with a candle mold cleaner, following the manufacturer's directions.

CANDLE WICKS

The wick must be carefully selected for the type of candle you are making to insure proper burning. Wicks come in four basic types:

Flat-braided wicks are best used in dipped taper candles. Always use this wick with braid Vs facing to the top of the candle. This prevents carbonized balls from forming on the end of the wick.

Paper-core wicks are used for container candles, votives, and tea lights. However, they have a tendency to smoke more than other wicks.

Square-braided wicks are sturdier than flat-braided wicks and are used in pillar candles.

Wire-core wicks are primarily used for container candles, votives, and tea lights. Some wire-core wicks come pretabbed and need only to be anchored to the bottom of the mold or container. The metal core in these wicks is typically made from zinc, although you may find some with lead, which may contain hazardous fumes when burned.

Wicks can be purchased primed or unprimed. Wicks that have been primed work best. If the wicks are not primed, you can prime the wick by soaking it in melted wax for five minutes. Remove the wick from the wax and lay it straight on waxed paper and allow to cool.

Size (or diameter) and length of the wick is usually determined by the diameter and length of the candle. Use a small-size wick for candles up to two inches

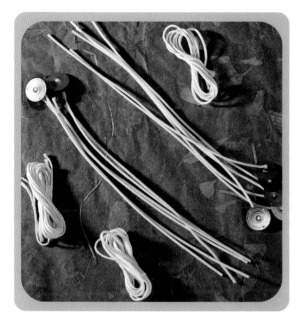

diameter of the candle or the wax from which it is formed. To achieve consistent results, keep notes about what works in the various candles you create.

CANDLE WICK TABS

Candle wick tabs are small metal disks to which the wicks may be attached for secure placement in a candle. A small amount of melted wax is poured into the bottom center of the mold. The disk is then centered in the wax and secured as the wax cools. The remainder of the wax is then poured into the mold.

CANDLE DYES

Candle dyes are available in liquid, solid, and powdered form. Add the dye a little at a time until desired color is achieved, following the manufacturer's directions.

Liquid dye is added to the melted wax and blends in quite easily.

Solid dye consists of concentrated color pigment and wax. The tablets or chips should be cut up, then added to the melted wax and stirred until fully melted.

Powdered candle pigments should be dissolved in warmed stearic acid before being mixed into melted wax. Powdered pigments are so concentrated that only a very small amount is required for coloring wax.

in diameter, a medium-size wick for 2- to 3-inch diameter candles, and a large-size wick for 3- to 4-inch diameter candles. A smaller-size wick may be used for votive candles.

The length of the wick is determined by measuring the height of the mold and adding two inches. In some cases, you may wish to leave a longer wick that can be knotted or embellished with beads and charms for a more decorative effect or for gift giving.

Wick size is also determined by the type of wax that is used in the candle. Long-burning wax, such as beeswax or paraffin with hardening additives, will require a larger-size wick. Often, a poorly burning candle is caused by selecting the wrong wick for the

ADDING FRAGRANCE OILS

Citrus oils may cause a flare-up in burning candles, with orange being the worst. When using these oils, they should be included in quantities of less than 5%. It is a good idea to test the citrus oils in sample candles.

Fragrance oils made from spices are as intense and powerful as the spices themselves and should always be used sparingly. They may cause the flame to sputter as well as overwhelm the sense of smell. Never use ground spice in the core of the candle because it may congest the wick.

Several of the more expensive essential oils, such as jasmine, rose, and sandalwood, are often sold in dilution or as a synthetically manufactured fragrance. Always purchase oils from a trusted source. Because of their expense, these oils are best used in small amounts in a candle or in an aroma lamp or diffuser.

WAX TEMPERATURES FOR CANDLEMAKING

- ACRYLIC MOLDS   180°F to 210°F
- BAYBERRY CANDLES   130°F
- CONTAINER CANDLES   160°F to 165°F
- DIPPING CANDLES   155°F to 160°F
- GLASS MOLDS   170°F to 200°F
- METAL MOLDS   180°F to 210°F
- RUBBER MOLDS   160°F to 180°F
- TAPER CANDLES   158°F
- TEAR-AWAY MOLDS   160°F

PREPARATION TIPS

Clear work area and cover surface with waxed paper to protect against drips or spills. Avoid wax drips or spills on burners or range at all costs because of the flammability of the wax. Never pour wax into a pan sitting on a burner.

Always melt wax in a double boiler or water bath. Never melt wax and leave it unattended over direct heat. Wax will self-ignite when overheated.

Candles should always be left to cure at least twenty-four hours before being lit.

Do not extinguish burning wax with water. Instead, smother the flames with baking soda, a towel, a pot lid, or a blanket.

Always clean up spills as they happen. Melted wax, collecting in the proximity of a burner, is a fire hazard.

Never pour wax down a drain because it will clog the drain.

If hot wax is spilled on skin, do not wipe it off. Rinse the skin under cold water and allow the wax to set and then scrape it off.

Remove uncolored wax from fabric by allowing it to harden and then scraping it off; or place a paper napkin or facial tissue over and under the wax spot, and iron it with a hot iron. If the wax is dyed, it would be better to have the fabric professionally dry-cleaned so the color is not set into the fabric by the heat of the iron.

Use pots and containers that are dedicated to candlemaking. Retire these items from food use. Consider purchasing pots and containers from thrift stores or at garage sales.

Waxy residues on pots and pans can be removed by placing the container in the fridge or freezer until the wax becomes loose and is easily removed. See also section on Cleaning Molds.

Save all candle leftovers because they are valuable raw materials and be can remelted and used again. Let your friends know that you collect leftover candle odds and ends.

Always be aware of flowers, ribbons, and other decorative items that are in close proximity of a candle flame.

Never leave a candle unattended.

Va-Va-Voom Vanilla Candle

Vanilla has a base note with a very mild and soothing scent. The smooth vanilla scent is one that is spiritually warming and calming. Derived from an orchid seedpod and used in tincture form to flavor, vanilla offers a rich, grounding fragrance to any blend. It relaxes and revitalizes the mind and body and diffuses irritability and angst. Vanilla also calms nerves and reduces muscle tension. This scent blends well with clove, sandalwood, and vetiver.

MATERIALS

- mold-blend wax
- round pillar mold
- square-braided wick, primed
- 1 drop sandalwood fragrance oil
- 3 drops vanilla fragrance oil

1. Place wax chunks in top pan of double boiler. Place double boiler on range and heat water in bottom pan to boiling point. As wax begins to melt and pool, place thermometer into wax but do not allow it to touch bottom of pan. Reduce heat to medium low so that water continues at a gentle boil. Do not allow water to boil dry.

2. While wax is melting, prepare mold by lightly coating inside of mold with mold release.

3. Cut wick to desired length. Thread wick through hole in bottom of mold. Cover hole and secure end of wick to outside of mold with mold sealer to prevent leakage. Pull wick straight up through mold. Place metal skewer on top of mold. Make certain that wick is centered and taut, then wrap wick around metal skewer.

4. Pour melted wax into metal pouring pot.

5. Add fragrance oils to wax when ready to pour into mold. Stir fragrance oils to distribute scents. If wax is too hot, some of the oil may dissipate and some scent may be lost.

6. Slowly pour melted wax into mold until mold is 90% full. Pour excess wax into a small container and set aside. Lightly tap mold to release any air bubbles. Allow wax to cool. As wax cools, an indentation may form around the wick.

7. Place small container of wax into water of bottom pan of double boiler and remelt wax. Fill indentation with melted wax. Fill only to top of indentation to avoid making a line around outside edge of the candle. Allow wax to cool.

8. Remove skewer and mold sealer from ends of mold. Tip mold upside down and candle should slide out on its own. If candle does not slide out, place mold in freezer for 5 to 10 minutes. Remove mold from freezer and slide candle out.

9. Trim bottom end of wick flush with candle. Trim top of wick to ½ inch. For a decorative effect, wick may be left longer.

Lavish Lavender Candle

Lavender has a middle note with a floral scent that is clean and fresh. It is also one of the most versatile essential oils. Lavender has been used for centuries for scenting linens and baths. It can be used topically as well as a scent and washes away physical and mental impurities. In addition to uplifting spirits and relieving stress, lavender disinfects, enhances sleep, loosens congestion, reduces muscle tension, reduces pain, repels insects, and soothes and heals skin. Lavender blends well with chamomile, citrus, lemongrass, and rose scents.

MATERIALS
- mold-blend wax
- round pillar mold
  (larger than unscented pillar candle)
- unscented white pillar candle
- 2 drops clary sage fragrance oil
- 4 drops lavender fragrance oil
- 4 drops lemon fragrance oil
- glass marbles or pebbles

1. Place wax chunks in top pan of double boiler. Place double boiler on range and heat water in bottom pan to boiling point. As wax begins to melt and pool, place thermometer into wax but do not allow it to touch bottom of pan. Reduce heat to medium low so that water continues at a gentle boil. Do not allow water to boil dry.

2. While wax is melting, prepare mold by lightly coating inside of mold with mold release.

3. Light the white pillar candle and allow a pool of wax to form at the base of the wick. Blow out the flame and add 1 drop clary sage fragrance oil, 2 drops lavender fragrance oil, and 2 drops lemon fragrance oil to the pool of wax. As the wax cools, the oils are absorbed into the candle.

4. Center pillar candle in mold.

5. Drop glass marbles into mold until half the height of candle is covered.

6. Pour melted wax into metal pouring pot.

7. Add remaining fragrance oils to melted wax and stir.

8. Slowly pour melted wax over marbles until mold is 90% full. Pour excess wax into a small container. Allow wax in mold and container to set.

9. Remelt excess wax and fill mold with melted wax. Do not pour melted wax above top edge of pillar candle. Allow wax to set.

10. Tip mold upside down and remove candle.

11. Using a heat gun, melt away some wax to expose some glass marbles partially, giving a decorative uneven finish.

Tangy Citrus Candle

Orange has a top note with a fresh citrus scent. This scent relaxes, balances, and heals. Orange is used to uplift and energize the mind and spirit; it can help to ease feelings of depression and hopelessness. It is a cheerful scent that will have a powerful effect on the spirit's well-being, as well as reducing muscle tension and relieving anxiety and fear. Orange blends well with cinnamon, cypress, juniper, and sandalwood.

MATERIALS

- mold-blend wax
- tall triangular pillar mold
- square-braided wick, primed
- orange candle dye
- 5 drops cedarwood fragrance oil
- 3 drops orange fragrance oil
- 1 drop ylang-ylang fragrance oil
- seashells
- string

1. Place wax chunks in top pan of double boiler. Place double boiler on range and heat water in bottom pan to boiling point. As wax begins to melt and pool, place thermometer into wax but do not allow it to touch bottom of pan. Reduce heat to medium low so that water continues at a gentle boil. Do not allow water to boil dry.

2. While wax is melting, prepare mold by lightly coating inside of mold with mold release.

3. Cut wick to desired length. Thread wick through hole in bottom of mold. Cover hole and secure end of wick to outside of mold with mold sealer to prevent leakage. Pull wick straight up through mold. Place metal skewer on top of mold. Make certain that wick is centered and taut, then wrap wick around metal skewer.

4. Pour melted wax into metal pouring pot.

5. Add candle dye and stir until color is evenly distributed.

6. Add fragrance oils to wax when ready to pour into mold. Stir fragrance oils to distribute scents. If wax is too hot, some of the oil may dissipate and some scent may be lost.

7. Slowly pour melted wax into mold until mold is 90% full. Pour excess wax into a small container and set aside. Lightly tap mold to release any air bubbles. Allow wax to cool. As wax cools, an indentation may form around the wick.

8. Place small container of wax into water of bottom pan of double boiler and remelt wax. Fill indentation with melted wax. Fill only to top of indentation to avoid making a line around outside edge of the candle. Allow wax to cool.

9. Remove skewer and mold sealer from ends of mold. Tip mold upside down and candle should slide out on its own. If candle does not slide out, place mold in freezer for 5 to 10 minutes. Remove mold from freezer and slide candle out.

10. Wrap candle with string, starting at bottom of candle and wrapping in a spiraling motion toward top. Secure ends with small amount of melted wax.

11. Press seashells into sides of candle at top, using melted wax to secure.

12. Pour remaining melted wax into dipping vat. Holding candle by wick at top, dip into wax. Hold for a few seconds and lift out and allow to cool.

13. Trim bottom end of wick flush with candle. Trim top of wick to ½ inch. For a decorative effect, the wick may be left longer.

Stress-Bustin' Sage Candle

Clary sage has a top to middle note with a fresh, sweet scent. This scent is one that calms, eases pain, and reduces tension and stress. It warms and stimulates the body, and promotes restful sleep. Clary sage has also been noted for its uplifting and regenerating effect on the mind and body. This harmonious and herbaceous oil will encourage joy and relaxation, bringing a sense of peaceful bliss and balance to those who experience it. Clary sage blends well with geranium, jasmine, lavender, and orange.

MATERIALS
- mold-blend wax
- triangle pillar mold
- square-braided wick, primed
- green candle dye
- brown candle dye
- ivory candle dye
- 3 drops cedarwood fragrance oil
- 3 drops clary sage fragrance oil
- 6 drops grapefruit fragrance oil

1. Place wax chunks in top pan of double boiler. Place double boiler on range and heat water in bottom pan to boiling point. As wax begins to melt and pool, place thermometer into wax but do not allow it to touch bottom of pan. Reduce heat to medium low so that water continues at a gentle boil. Do not allow water to boil dry.

2. While wax is melting, prepare mold by lightly coating inside of mold with mold release.

3. Cut wick to desired length. Thread wick through hole in bottom of mold. Cover hole and secure end of wick to outside of mold with mold sealer to prevent leakage. Pull wick straight up through mold. Place metal skewer on top of mold. Make certain that wick is centered and taut, then wrap wick around metal skewer.

4. Pour ⅓ of melted wax into metal pouring pot.

5. Add green dye to melted wax and stir until color is evenly distributed.

6. Add 1 drop cedarwood fragrance oil, 1 drop clary sage fragrance oil, and 2 drops grapefruit fragrance oil to melted wax in metal pouring pot. Stir fragrance oils to distribute scents. If wax is too hot, some of the oil may dissipate and some scent may be lost.

7. Pour melted wax into mold until one-third full. Pour excess wax into a small container and set aside. Lightly tap mold to release any air bubbles. Allow wax to cool. As wax cools, an indentation may form around the wick.

8. Pour one-half of remaining melted wax into metal pouring pot and add brown dye. Stir until color is evenly distributed.

9. Add 1 drop cedarwood fragrance oil, 1 drop clary sage fragrance oil, and 2 drops grapefruit fragrance oil to melted wax in metal pouring pot. Stir fragrance oils to distribute scents.

10. Pour melted wax into mold until two-thirds full. Pour excess wax into a small container and set aside. Lightly tap mold to release any air bubbles. Allow wax to cool. As wax cools, an indentation may form around the wick.

11. Pour remaining melted wax into metal pouring pot and add ivory dye. Stir until color is evenly distributed.

12. Add remaining fragrance oils to melted wax. Stir fragrance oils to distribute scents.

13. Pour melted wax into mold until completely full. Pour excess wax into a small container and set

aside. Lightly tap mold to release any air bubbles. Allow wax to cool. As wax cools, an indentation may form around the wick.

14. Remelt excess ivory wax and fill indentation. Allow wax to set.

15. Remove skewer and mold sealer from ends of mold. Tip mold upside down and candle should slide out on its own. If candle does not slide out, place mold in freezer for 5 to 10 minutes. Remove mold from freezer and slide candle out.

16. Trim bottom end of wick flush with candle. Trim top of wick to ½ inch. For a decorative effect, wick may be left longer.

Off-Limits Citronella & Bay Leaf Candle

Citronella has a middle-to-top note with a sweet citrus like scent. Citronella is known for its powerful insect-repelling qualities and is favored for use in outdoor garden candles. Citronella's fresh and fruity properties help ease headaches and eliminate fatigue. Citronella blends well with clove, jasmine, juniper berry, nutmeg, and vetiver.

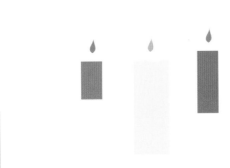

MATERIALS

- glue pen with sponge applicator
- bay leaves
- unscented ivory pillar candle, 3" by 6"
- taper-blend wax
- 1 drop citronella fragrance oil
- 2 drops lavender fragrance oil
- 2 drops lemongrass fragrance oil
- brown candle dye
- sage-green ribbon

1. Apply glue to backs of bay leaves. Adhere leaves to pillar candle. Leaves can be carefully repositioned if glue is still wet. Allow glue to dry.

2. Determine amount of taper-blend wax needed to fill dipping vat by filling vat with water and measuring the water. Three ounces of wax is equivalent to 3½ ounces of water.

3. Place wax chunks in top pan of double boiler. Place double boiler on range and heat water in bottom pan to boiling point. As wax begins to melt and pool, place thermometer into wax but do not allow it to touch bottom of pan. Reduce heat to medium low so that water continues at a gentle boil. Do not allow water to boil dry.

4. Light pillar candle and allow a pool of wax to form at the base of the wick. Blow out the flame and add the fragrance oils to the pool of wax. As the wax cools, the oils are absorbed into the candle.

5. Pour melted wax into dipping vat.

6. Add candle dye to melted wax and stir until color is evenly distributed.

7. Holding candle by wick, dip candle into dipping vat in one smooth motion. Repeat dipping two additional times. Place on waxed paper and allow wax to set.

8. Tie ribbon around candle.

Glowing Goddess Candle

Ginger has a middle-to-top note with an earthy and spicy aroma. It is lightening to the senses and warming to the soul. Ginger is believed to be high in yang energy by the Chinese. This spicy root is used for flavoring foods as well as health purposes, such as enhancing appetite, heightening senses, relieving stress, and soothing body and mind. Ginger blends well with lavender, lemon, lime, neroli, orange, and rosewood.

MATERIALS

- container-blend wax
- tea-light mold
- square-braided wick, primed
- white candle dye
- 2 drops ginger fragrance oil
- 2 drops lavender fragrance oil
- 1 drop neroli fragrance oil
- dried lavender

1. Place wax chunks in top pan of double boiler. Place double boiler on range and heat water in bottom pan to boiling point. As wax begins to melt and pool, place thermometer into wax but do not allow it to touch bottom of pan. Reduce heat to medium low so that water continues at a gentle boil. Do not allow water to boil dry.

2. While wax is melting, prepare mold by lightly coating inside of mold with mold release.

3. Cut wick to desired length. Thread wick through hole in bottom of mold. Cover hole and secure end of wick to outside of mold with mold sealer to prevent leakage. Pull wick straight up through mold. Place metal skewer on top of mold. Make certain that the wick is centered and taut, then wrap wick around metal skewer.

4. Pour melted wax into metal pouring pot.

5. Add candle dye and stir until color is evenly distributed.

6. Add fragrance oils to wax when ready to pour into mold. Stir fragrance oils to distribute scents. If wax is too hot, some of the oil may dissipate and some scent may be lost.

7. Place dried lavender into melted wax and stir.

8. Slowly pour melted wax into mold until mold is 90% full. Pour excess wax into a small container and set aside. Lightly tap mold to release any air bubbles. Allow wax to cool. As wax cools, an indentation may form around the wick.

9. Place small container of wax into water of bottom pan of double boiler and remelt wax. Fill indentation with melted wax. Fill only to top of indentation to avoid making a line around outside edge of candle. Allow wax to cool.

10. Remove skewer and mold sealer from ends of mold. Tip mold upside down and candle should slide out on its own. If candle does not slide out, place mold in freezer for 5 to.10 minutes. Remove mold from freezer and slide candle out.

11. Trim bottom end of wick flush with candle. Trim top of wick to ½ inch. For a decorative effect,

Nature Grrrl Candle

Cypress has a middle note with a fresh woodsy scent. The cypress scent is one that promotes mental concentration. It is also spiritually calming and refreshing. Cypress helps to regulate female hormones and has been known to reduce the appearance of cellulite. This warming and purifying aroma has the ability to rejuvenate and awaken the senses, to improve coping skills, relieve tension and stress, stimulate circulation, reduce muscle tension, and encourage restful sleep. Cypress blends well with juniper, lavender, lemon, and pine.

MATERIALS

- mold-blend wax
- round pillar mold
- square-braided wick, primed
- ivory candle dye
- 4 drops bergamot fragrance oil
- 5 drops cypress fragrance oil
- 2 drops petitgrain fragrance oil
- acrylic texture medium
- ground liquid iron
- rusting compound

1. Place wax chunks in top pan of double boiler. Place double boiler on range and heat water in bottom pan to boiling point. As wax begins to melt and pool, place thermometer into wax but do not allow it to touch bottom of pan. Reduce heat to medium low so that water continues at a gentle boil. Do not allow water to boil dry.

2. While wax is melting, prepare mold by lightly coating inside of mold with mold release.

3. Cut wick to desired length. Thread wick through hole in bottom of mold. Cover hole and secure end of wick to outside of mold with mold sealer to prevent leakage. Pull wick straight up through mold. Place metal skewer on top of mold. Make certain that wick is centered and taut, then wrap wick around metal skewer.

4. Pour melted wax into pouring pot.

5. Add candle dye and stir until color is evenly distributed.

6. Add fragrance oils to wax when ready to pour into mold. Stir fragrance oils to distribute scents. If wax is too hot, some of the oil may dissipate and some scent may be lost.

7. Slowly pour melted wax into mold until mold is 90% full. Pour excess wax into a small container and set aside. Lightly tap mold to release any air bubbles. Allow wax to cool. As wax cools, an indentation may form around the wick.

8. Place small container of wax into water of bottom pan of double boiler and remelt wax. Fill indentation with melted wax. Fill only to top of indentation to avoid making a line around outside edge of the candle. Allow wax to cool.

9. Remove skewer and mold sealer from ends of mold. Tip mold upside down and candle should slide out on its own. If candle does not slide out, place mold in freezer for 5 to 10 minutes. Remove mold from freezer and slide candle out.

10. Trim bottom end of wick flush with candle. Trim top of wick to ½ inch. For a decorative effect, wick may be left longer.

11. Using palette knife, daub acrylic texture medium onto candle to create a textured look, following manufacturer's directions. Allow to dry.

12. Paint with liquid iron, carefully filling in between spaces in texture medium. Allow to dry.

13. Apply rusting compound to textured area of candle, following manufacturer's directions. Allow to rust and dry.

Pep Up Peppermint Candle

Peppermint has a top note with a very fresh and minty scent. This scent is one that stimulates and energizes the mind and body. It awakens the senses and encourages mental clarity. It has been noted to relieve pain and alleviate cold and flu symptoms such as congestion, dizziness, fever, headaches, and vomiting. Peppermint's refreshing qualities will uplift and enlighten the mood, and increase strength while relieving fatigue. Peppermint blends well with clove, eucalyptus, lime, and marjoram.

MATERIALS
- mold-blend wax
- round pillar mold
- square-braided wick, primed
- light green wax chunks
- 2 drops lavender fragrance oil
- 4 drops peppermint fragrance oil
- 3 drops rosemary fragrance oil

1. Place mold-blend wax chunks in top pan of double boiler. Place double boiler on range and heat water in bottom pan to boiling point. As wax begins to melt and pool, place thermometer into wax but do not allow it to touch bottom of pan. Reduce heat to medium low so that water continues at a gentle boil. Do not allow water to boil dry.

2. While wax is melting, prepare mold by lightly coating inside of mold with mold release.

3. Cut wick to desired length. Thread wick through hole in bottom of mold. Cover hole and secure end of wick to outside of mold with mold sealer to prevent leakage. Pull wick straight up through mold. Place metal skewer on top of mold. Make certain that wick is centered and taut, then wrap wick around metal skewer.

4. Position green wax chunks in candle mold as desired.

5. Pour melted wax into metal pouring pot.

6. Add fragrance oils to wax when ready to pour into mold. Stir fragrance oils to distribute scents. If wax is too hot, some of the oil may dissipate and some scent may be lost.

7. Slowly pour melted wax into mold until mold is 90 percent full. Pour excess wax into a small container and set aside. Lightly tap mold to release any air bubbles. Allow wax to cool. As wax cools, an indentation may form around the wick.

8. Place a small container of wax into water of bottom pan of double boiler and remelt wax. Fill indentation with melted wax. Fill only to top of indentation to avoid making a line around outside edge of the candle. Allow wax to cool.

9. Remove skewer and mold sealer from ends of mold. Tip mold upside down and candle should slide out on its own. If candle does not slide out, place mold in freezer for 5 to 10 minutes. Remove mold from freezer and slide candle out.

10. Trim bottom end of wick flush with candle. Trim top of wick to ½ inch. For a decorative effect, wick may be left longer.

Find Your Center Candle

Juniper berry has a middle note that is energizing and balancing. In addition to repelling insects, it improves concentration and mental clarity. Combined with elements of feng shui, the Chinese art of harmonizing shapes and colors with spiritual forces, juniper berry makes an ideal candle for intuition and self-cultivation. The color green signifies wood. Wood promotes growth, new beginnings, freshness, and nurturing. The color purple signifies prosperity. Juniper berry blends well with bergamot, rosemary, and sandalwood.

MATERIALS

- mold-blend wax
- round pillar mold
- square-braided wick, primed
- green candle dye
- purple candle dye
- 6 drops juniper berry fragrance oil

1. Place wax chunks in top pan of double boiler. Place double boiler on range and heat water in bottom pan to boiling point. As wax begins to melt and pool, place thermometer into wax but do not allow it to touch bottom of pan. Reduce heat to medium low so that water continues at a gentle boil. Do not allow water to boil dry.

2. While wax is melting, prepare mold by lightly coating inside of mold with mold release.

3. Cut wick to desired length. Thread wick through hole in bottom of mold. Cover hole and secure end of wick to outside of mold with mold sealer to prevent leakage. Pull wick straight up through mold. Place metal skewer on top of mold. Make certain that wick is centered and taut, then wrap wick around metal skewer.

4. Pour two-thirds of melted wax into metal pouring pot.

5. Add purple candle dye to melted wax in metal pouring pot and stir until color is evenly distributed.

6. Add 4 drops of fragrance oil to melted purple wax when ready to pour into mold. Stir fragrance oil to distribute scents. If wax is too hot, some of the oil may dissipate and some scent may be lost.

7. Slowly pour melted purple wax into mold until two-thirds full. Allow wax to cool.

8. Pour remaining melted wax into metal pouring pot and add green candle dye. Stir until color is evenly distributed

9. Add remaining fragrance oil to melted green wax when ready to pour into mold. Stir fragrance oil to distribute scents.

10. Pour melted green wax into mold until mold is 90% full. Pour excess green wax into a small container and set aside. Lightly tap mold to release any air bubbles. Allow wax to cool. As wax cools, an indentation may form around the wick.

11. Remelt any excess green wax and fill indentation with melted wax. Allow wax to cool.

12. Remove skewer and mold sealer from ends of mold. Tip mold upside down and candle should slide out on its own. If candle does not slide out, place mold in freezer for 5 to 10 minutes. Remove mold from freezer and slide candle out.

13. Trim bottom end of wick flush with candle. Trim top of wick to ½ inch. For a decorative effect, wick may be left longer.

MATERIALS

- taper-blend wax
- green candle dye
- purple candle dye
- 4 drops cypress fragrance oil
- 2 drops ginger fragrance oil
- 4 drops spruce fragrance oil
- medium-sized pinecones

1. Place wax chunks in top pan of double boiler. Place double boiler on range and heat water in bottom pan to boiling point. As wax begins to melt and pool, place thermometer into wax but do not allow it to touch bottom of pan. Reduce heat to medium low so that water continues at a gentle boil. Do not allow water to boil dry.

2. Evenly pour melted wax into two dipping vats.

3. Add green candle dye to first vat and purple dye to the second. Stir both vats until color is evenly distributed.

4. When ready to dip pinecones, add 2 drops cypress fragrance oil, 1 drop ginger fragrance oil, and 2 drops spruce fragrance oil to green wax. Add remaining fragrance oils to purple wax. Stir fragrance oils to distribute scents. If wax is too hot, some of the oil may dissipate and some scent may be lost.

Festive Pinecone Candles

Light these pinecone candles in the fireplace to add the aroma of a pine forest to the room. Spruce has a middle note with a woody scent that is fresh with a slight hint of fruit. It is calming to the nervous system and encourages communication as well as sleep. Spruce blends well with lemon, peppermint, and spearmint.

5. Using slotted spoon, dip pinecone into green wax and remove.

6. Place pinecone onto waxed paper. Allow to cool.

7. Alternating between green and purple waxes, repeat for remaining pinecones.

Looking-Good Lemon Candle

Lemon has a top note with a fresh citrus scent. The soothing scent is cooling, balancing, improves mental clarity, and helps to relieve fatigue. This essential oil is extracted from the peel and lends a citrusy scent to any blend. Lemon can be antiseptic and is good for building up the immune system. Lemon blends well with cedarwood, eucalyptus, fennel, juniper, and ylang-ylang.

MATERIALS
- mold-blend wax
- round pillar mold, 3 inches in diameter
- yellow wax chunks
- square-braided wick, primed
- 1 drop lavender fragrance oil
- 3 drops lemon fragrance oil
- 1 drop lemongrass fragrance oil
- fruit beads

1. Place mold-blend wax chunks in top pan of double boiler. Place double boiler on range and heat water in bottom pan to boiling point. As wax begins to melt and pool, place thermometer into wax but do not allow it to touch bottom of pan. Reduce heat to medium low so that water continues at a gentle boil. Do not allow water to boil dry.

2. While wax is melting, prepare mold by lightly coating inside of mold with mold release.

3. Cut wick to desired length and add 3 inches. Thread wick through hole in bottom of mold. Cover hole and secure end of wick to outside of mold with mold sealer to prevent leakage. Pull wick straight up through mold. Place metal skewer on top of mold. Make certain that wick is centered and taut, then wrap wick around metal skewer.

4. Position several yellow wax chunks in candle mold as desired.

5. Pour melted wax into metal pouring pot.

6. Add fragrance oils to melted wax when ready to pour into mold. Stir fragrance oils to distribute scents. If wax is too hot, some of the oil may dissipate and some scent may be lost.

7. Slowly pour melted wax into mold until mold is 90% full. Pour excess wax into a small container and set aside. Lightly tap mold to release any air bubbles. Allow wax to cool.

8. Place additional yellow wax chunks into mold, allowing them to protrude from melted wax. Allow wax to cool.

9. Remove skewer and mold sealer from ends of mold. Tip mold upside down and candle should slide out on its own. If candle does not slide out, place mold in freezer for 5 to 10 minutes. Remove mold from freezer and slide candle out.

10. Trim bottom end of wick flush with candle. Trim top of wick to remain long enough to thread on fruit beads for decoration. Fruit beads must be removed and wick trimmed before burning.

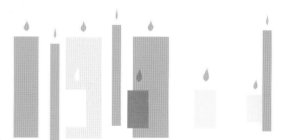

Bayberry Babe Pillar Candles

Bayberry wax has a soothing and fragrant aroma that does not require the addition of other scents. It is created by boiling the scented berries of the bayberry bush and skimming off the wax, which retains the naturally occurring scent. Bayberry is not a true essential oil and it is usually found as a synthetic fragrance or bayberry essence in a carrier oil. The berries and bark of the bayberry bush have astringent properties and are used to aid intestinal problems and jaundice. It has been a tradition since colonial days to burn a bayberry candle down to a nub on New Year's Eve to bring health and happiness for the coming year.

MATERIALS
- bayberry wax
- 1½" by 10" stick mold
- square-braided wick, primed

1. Place wax chunks in top pan of double boiler. Place double boiler on range and heat water in bottom pan to boiling point. As wax begins to melt and pool, place thermometer into wax but do not allow it to touch bottom of pan. Reduce heat to medium low so that water continues at a gentle boil. Do not allow water to boil dry.

2. While wax is melting, prepare mold by lightly coating inside of mold with mold release.

3. Cut wick to desired length. Thread wick through hole in bottom of mold. Cover hole and secure end of wick to outside of mold with mold sealer to prevent leakage. Pull wick straight up through

mold. Place metal skewer on top of mold. Make certain that wick is centered and taut, then wrap wick around metal skewer.

4. Slowly pour melted wax into mold until mold is 90% full. Pour excess wax into a small container and set aside. Lightly tap mold to release any air bubbles. Allow wax to cool. As wax cools, an indentation may form around the wick.

5. Place small container of wax into water of bottom pan of double boiler and remelt wax. Fill indentation with melted wax. Fill only to top of indentation to avoid making a line around outside edge of the candle. Allow wax to cool.

6. Remove skewer and mold sealer from ends of mold. Tip mold upside down and candle should slide out on its own. If candle does not slide out, place mold in freezer for 5 to 10 minutes. Remove mold from freezer and slide candle out.

7. Trim bottom end of wick flush with candle. Trim top of wick to ½ inch. For a decorative effect, wick may be left longer.

TIP: Bayberry wax may be combined with beeswax in a ratio of 8:2.

Hanky-Panky Candle

In India, this fragrant, sensual flower has been called "Queen of the Night." Renowned as an "aphrodisiac, jasmine will help you relax by inspiring one of life's most soothing rewards—love. It warms emotions and restores positive feelings. Jasmine can also be used as an antidepressant, to reduce muscle tension, and to calm the body and mind.

MATERIALS
- mold-blend wax
- round pillar mold
- square-braided wick, primed
- red candle dye
- 2 drops jasmine fragrance oil
- 1 drop rose fragrance oil
- 1 drop ylang-ylang fragrance oil

1. To create a rustic look on outside of candle, place mold into freezer for 30 minutes.

2. Place wax chunks in top pan of double boiler. Place double boiler on range and heat water in bottom pan to boiling point. As wax begins to melt and pool, place thermometer into wax but do not allow it to touch bottom of pan. Reduce heat to medium low so that water continues at a gentle boil. Do not allow water to boil dry.

3. While wax is melting, prepare mold by lightly coating inside of mold with mold release.

4. Cut wick to desired length. Thread wick through hole in bottom of mold. Cover hole and secure end of wick to outside of mold with mold sealer to prevent leakage. Pull wick straight up through mold. Place metal skewer on top of mold. Make certain that wick is centered and taut, then wrap wick around metal skewer.

5. Pour melted wax into metal pouring pot.

6. Add candle dye and stir until color is evenly distributed.

7. Add fragrance oils to wax when ready to pour into mold. Stir fragrance oils to distribute scents. If wax is too hot, some of the oil may dissipate and some scent may be lost.

8. Slowly pour melted wax into mold until mold is 90% full. Pour excess wax into a small container and set aside. Lightly tap mold to release any air bubbles. Allow wax to cool. As wax cools, an indentation may form around the wick.

9. Place small container of wax into water of bottom pan of double boiler and remelt wax. Fill indentation with melted wax. Fill only to top of indentation to avoid making a line around outside edge of candle. Allow wax to cool.

10. Remove skewer and mold sealer from ends of mold. Tip mold upside down and candle should slide out on its own. If candle does not slide out,

place mold in freezer for 5 to 10 minutes. Remove mold from freezer and slide candle out.

11. Using craft knife or other carving tool, carve hearts of various sizes onto outside of candle in random pattern. Using flat edge of craft knife, carefully scrape wax away from inside of heart shapes. A soft cloth may be used to buff out scrape marks from inside hearts.

12. Trim bottom end of wick flush with candle. Trim top of wick to ½ inch. For a decorative effect, wick may be left longer.

Baby, Light My Fire!

Clove, a warming and stimulating fragrance, is also considered an aphrodisiac. Because of its spicy scent, clove should be used sparingly in any blend. In addition to uplifting the spirit, relieving pain, and improving mental clarity and memory, clove oil disinfects, improves digestion, loosens congestion, and reduces fatigue.

MATERIALS

- mold-blend wax
- large oval mold
- unscented white round pillar candle, smaller than mold
- assorted dried foliage
- gold candle dye
- 3 drops clove fragrance oil
- 4 drops orange fragrance oil
- 5 drops vanilla fragrance oil
- dried orange slices
- pepper berries
- large flat leaves

1. Place wax chunks in top pan of double boiler. Place double boiler on range and heat water in bottom pan to boiling point. As wax begins to melt and pool, place thermometer into wax but do not allow it to touch bottom of pan. Reduce heat to medium low so that water continues at a gentle boil. Do not allow water to boil dry.

2. While wax is melting, prepare mold by lightly coating inside of mold with mold release.

3. Light the white pillar candle and allow a pool of wax to form at the base of the wick. Blow out the flame and add 1 drop clove fragrance oil, 2 drops orange fragrance oil, and 2 drops vanilla fragrance oil to the pool of wax. As the wax cools, the oils are absorbed into the candle.

4. Center pillar candle in mold.

5. Arrange the dried orange slices, pepper berries, leaves, and assorted foliage around the entire pillar candle in the mold.

6. Pour melted wax into metal pouring pot.

7. Add candle dye and remaining fragrance oils to melted wax. Stir to distribute evenly.

8. Slowly pour melted wax over botanicals until mold is 90% full. Pour excess wax into a small container and allow to set.

9. Using a dowel (or chopstick), rearrange any botanicals that may float out of place. Allow wax to set.

10. Place small container of wax into water of bottom pan of double boiler and remelt wax. Fill indentation with melted wax. Fill only to top of indentation to avoid making a line around outside edge of candle. Allow wax to cool.

11. Tip mold upside down and remove candle.

Wild Thing Candle

The scent of pine uplifts, energizes, and refreshes. Inhaling the vapors of pine will help your breathing and purify your body. Pine can also be used to calm the nerves, enhance sleep and relieve anxiety and stress.

MATERIALS
- mold-blend wax
- tall triangular pillar mold
- square-braided wick, primed
- yellow candle dye
- 1 drop orange fragrance oil
- 1 drop pine fragrance oil
- 1 drop spruce fragrance oil

1. To create a rustic look on outside of candle, place mold into freezer for 30 minutes.

2. Place wax chunks in top pan of double boiler. Place double boiler on range and heat water in bottom pan to boiling point. As wax begins to melt and pool, place thermometer into wax but do not allow it to touch bottom of pan. Reduce heat to medium low so that water continues at a gentle boil. Do not allow water to boil dry.

3. While wax is melting, prepare mold by lightly coating inside of mold with mold release.

4. Cut wick to desired length. Thread wick through hole in bottom of mold. Cover hole and secure end of wick to outside of mold with mold sealer to prevent leakage. Pull wick straight up through mold. Place metal skewer on top of mold. Make certain that wick is centered and taut, then wrap wick around metal skewer.

5. Pour melted wax into metal pouring pot.

6. Add candle dye and stir until color is evenly distributed.

7. Add fragrance oils to wax when ready to pour into mold. Stir fragrance oils to distribute scents. If wax is too hot, some of the oil may dissipate and some scent may be lost.

8. Slowly pour melted wax into mold until mold is 90% full. Pour excess wax into a small container and set aside. Lightly tap mold to release any air bubbles. Allow wax to cool. As wax cools, an indentation may form around the wick.

9. Place small container of wax into water of bottom pan of double boiler and remelt wax. Fill indentation with melted wax. Fill only to top of indentation to avoid making a line around outside edge of candle. Allow wax to cool.

10. Remove skewer and mold sealer from ends of mold. Tip mold upside down and candle should slide out on its own. If candle does not slide out, place mold in freezer for 5 to 10 minutes. Remove mold from freezer and slide candle out.

11. Trim bottom end of wick flush with candle. Trim top of wick to ½ inch. For a decorative effect, wick may be left longer.

TIP: For a festive arrangement, you can make several of these candles, varying the size, shape, and color. After melting the wax, divide it into several containers and color and pour each separately.

Too Hot to Handle Candle

Cinnamon has a warm, spicy scent that is reminiscent of holidays and the aroma of freshly baked cookies. It stimulates, strengthens, and warms the mind, spirit, and body in addition to purifying the air in a room. Cinnamon is also an aphrodisiac, and it enhances mental clarity, improves digestion, reduces pain, and stimulates circulation.

MATERIALS

- mold-blend wax, plus one-half as much more for whipping
- round pillar mold, 4-inch diameter
- square-braided wick, primed
- red candle dye
- 4 drops cedarwood fragrance oil
- 1 drop cinnamon fragrance oil
- 1 drop clove fragrance oil
- 3 drops orange fragrance oil

1. Place wax chunks in top pan of double boiler. Place double boiler on range and heat water in bottom pan to boiling point. As wax begins to melt and pool, place thermometer into wax but do not allow it to touch bottom of pan. Reduce heat to medium low so that water continues at a gentle boil. Do not allow water to boil dry.

2. While wax is melting, prepare mold by lightly coating inside of mold with mold release.

3. Cut wick to desired length. Thread wick through hole in bottom of mold. Cover hole and secure end of wick to outside of mold with mold sealer to prevent leakage. Pull wick straight up through mold. Place metal skewer on top of mold. Make certain that wick is centered and taut, then wrap wick around metal skewer.

4. Pour melted wax into metal pouring pot.

5. Add candle dye and stir until color is evenly distributed.

6. Add fragrance oils to wax when ready to pour into mold. Stir fragrance oils to distribute scents. If wax is too hot, some of the oil may dissipate and some scent may be lost.

7. Slowly pour melted wax into mold until mold is 90% full. Pour excess wax into a small container and set aside. Lightly tap mold to release any air bubbles. Allow wax to cool. As wax cools, an indentation may form around the wick.

8. Place small container of wax into water of bottom pan of double boiler and remelt wax. Fill indentation with melted wax. Fill only to top of indentation to avoid making a line around outside edge of candle. Allow wax to cool.

9. Remove skewer and mold sealer from ends of mold. Tip mold upside down and candle should slide out on its own. If candle does not slide out, place mold in freezer for 5 to 10 minutes. Remove mold from freezer and slide candle out.

10. Allow the remaining melted wax to cool until thin film appears on wax surface.

11. Using fork, whip wax for about 5 to 10 minutes, until thick and foamy.

12. Warm sides of pillar candle with hair dryer.

13. Using spatula, apply a thick coat of whipped wax to top and sides of candle. (Note: Begin in one spot and cover that area completely before moving on to the next.) Allow wax to set.

14. Trim bottom end of wick flush with candle. Trim top of wick to ½ inch. For a decorative effect, wick may be left longer.

Perfect Thyme-ing Candle

Thyme has a top note that is herbaceous and fresh. It warms and purifies the body as well as relaxing tight muscles. Thyme also enhances mood, improves digestion, and reduces pain. Combined with elements of feng shui—the Chinese art of arrangement—thyme is an ideal scent to nurture communication and judgment. In this candle, the color red signifies fire, which promotes action and motivation. The color yellow signifies earth, a symbol for security and stability.

MATERIALS
- mold-blend wax
- round pillar mold
- square-braided wick, primed
- red candle dye
- yellow candle dye
- 6 drops thyme fragrance oil

1. Place wax chunks in top pan of double boiler. Place double boiler on range and heat water in bottom pan to boiling point. As wax begins to melt and pool, place thermometer into wax but do not allow it to touch bottom of pan. Reduce heat to medium low so that water continues at a gentle boil. Do not allow water to boil dry.

2. While wax is melting, prepare mold by lightly coating inside of mold with mold release.

3. Cut wick to desired length. Thread wick through hole in bottom of mold. Cover hole and secure end of wick to outside of mold with mold sealer to prevent leakage. Pull wick straight up through mold. Place metal skewer on top of mold. Make certain that wick is centered and taut, then wrap wick around metal skewer.

4. Pour two-thirds of melted wax into metal pouring pot.

5. Add red dye to melted wax in pouring pot and stir until color is evenly distributed.

6. Add 4 drops of fragrance oil to melted wax in metal pouring pot. Stir fragrance oil to distribute scent. If wax is too hot, some of the oil may dissipate and some scent may be lost.

7. Slowly pour melted red wax into mold until two-thirds full. Pour excess red wax into a small container and set aside. Lightly tap mold to release any air bubbles. Allow wax to cool. As wax cools, an indentation may form around the wick.

8. Pour remaining uncolored melted wax into metal pouring pot and add yellow dye. Stir until color is evenly distributed.

9. Add remaining fragrance oil to melted yellow wax in pouring pot. Stir fragrance oil to distribute scent.

10. Pour melted yellow wax into mold until full. Pour excess yellow wax into a small container and set aside. Lightly tap mold to release any air bubbles. Allow wax to cool. As wax cools, an indentation may form around the wick.

11. Remelt any excess yellow wax and fill indentation. Allow wax to set.

12. Remove skewer and mold sealer from ends of mold. Tip mold upside down and candle should slide out on its own. If candle does not slide out, place mold in freezer for 5 to 10 minutes. Remove mold from freezer and slide candle out.

13. Trim bottom end of wick flush with candle. Trim top of wick to ½ inch. For a decorative effect, wick may be left longer.

Bathing Beauty

CHAPTER 3

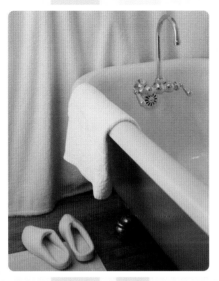

A bath is perhaps the easiest, quickest, and cheapest way to make you look and feel gorgeous. And who can resist a bargain, especially a bargain that makes you look good? While a bath alone is guaranteed to relax you, by adding fragrance oils to your bath, you enhance the water's soothing, calming, and invigorating effects. While you breathe in the wonderful aromas, your skin is absorbing the healing properties of the fragrance oils. The combination of water and fragrance oils will help you glow inside as well as outside! This chapter first explains how to dilute fragrance oils with a carrier oil, a very necessary step. Then based on your mood, pick any of the fabulous bath blends, oils, or salts, and use them to feel as beautiful as a tranquill mermaid, calmly swimming through your ocean of life.

Bath Secrets

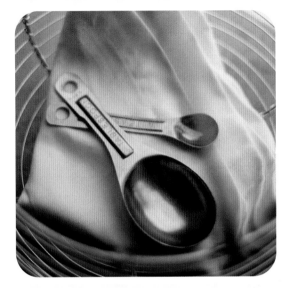

CARRIER OILS

The majority of essential oils must be mixed with a carrier oil in order to be used directly on the skin. Because of their concentrated nature, the oils can irritate the skin if not diluted. Carrier or base oils are vegetable oils used for diluting pure essential oils. They are more than just vehicles for essential oils, however, as they often have healthy benefits of their own. Combined with essential oils, they will add considerably to the dynamics of the blend used in various preparations, such as bath, body, facial, and massage products. By adding the aromatic essences to one or a combination of recommended base oils, the essential oils in these beauty products can be spread over a larger area of the body, allowing the beneficial properties of the oils to penetrate a wider surface of skin. As a general rule, the ratio of essential oil to carrier/base oil for adults should fall between 2 and 3%. This translates to about forty to sixty drops to a 4-ounce bottle of carrier oil, twenty to thirty drops to a 2-ounce bottle, and six to nine drops for a ½ ounce bottle.

You may choose from a variety of bases, depending on individual needs and preferences. Select a base that is as high in quality as your pure essential oil. Never use a synthetic or mineral oil base; these do not penetrate the skin nor provide any therapeutic benefits. Base oils are available in quantities from 1 ounce on up. Buy them in the smallest quantity available if you cannot use the oils within three to six months. Oils kept longer will become rancid. Usually 1 to 4 ounces is a good amount to have on hand to start. After using them for a while, you will discover which you prefer and can buy amounts accordingly.

Carrier oils can be obtained from health food stores, herb shops, or through mail order, and should be unrefined cold-pressed oils, rather than heat-treated or refined oils. Even without the addition of essential oils, these vegetable oils are health treatments in themselves.

TYPES OF CARRIER OILS

Following is a description of the carrier oils you may want to have on hand for experimentation. Most recipes in this book can be made with any of the carrier oils. Choose the oil or combination of oils that best suits your skin type.

APRICOT KERNEL OIL:

Good for all types of skin, especially useful on sensitive and aging skin. One of the lightest oils to use. Very good for a facial oil.

AVOCADO OIL:

A nutrient-rich base oil with a high content of vitamins, protein, lecithin, and essential fatty acids. Beneficial for all skin types, but especially for mature, wrinkled, dry, and itchy skin.

EVENING PRIMROSE OIL:

Expensive but a wonderful oil for skin care because it increases and protects skin cell function and works as a skin rejuvenator. The oil can quickly become rancid so it should be refrigerated. A small portion can be added to skin creams and lotions to increase their effectiveness. It is useful for dry skin, eczema, and psoriasis.

GRAPESEED OIL:

A nice, light, nonodoriferous oil. Makes a nice massage oil by itself or combined with sweet almond oil. Easily absorbed by the skin, suiting all skin types.

JOJOBA OIL:

Nourishing to the skin and hair. An oil rich in vitamin E that can be used alone or mixed with other base oils. Suitable for all skin types. Actually, it is a wax so it is unlikely to become rancid (unlike most of the other vegetable oils). Contains antibacterial properties, which makes it a very good treatment for acne.

SWEET ALMOND OIL:

Great base for massage, bath, body, and skin-care products because it is so nourishing to the skin. Contains a variety of vitamins and minerals, most notably vitamin D. Scentless. Suitable for all skin types, especially dry or irritated skin.

WHEAT GERM OIL:

An antioxidant oil; adding a small proportion to a basic mix (such as 1 tablespoon to every 2 ounces of massage or body oil) will retain the freshness of the blend and help extend the product's shelf life. High in vitamins E, A, and B as well as minerals and protein. Particularly beneficial to dry and mature skin. Also helps heal scar tissue, soothe burns, and smooth stretch marks. This oil should not be used on people who have wheat intolerance.

EQUIPMENT NEEDED FOR MAKING PRODUCTS

Once you have acquired essential oils and carrier/base oils, plus the ingredients listed for each product, the rest of the equipment needed can usually be found in your kitchen. Below is a list of equipment you will need:

- GLASS BOTTLES (DARK FOR MASSAGE OILS AND OTHER BLENDS) WITH DROPPER TOPS: for containing your mixtures. Use bottles with dropper tops instead of rubber tops, as the rubber will eventually break down. You may use clear glass or plastic bottles and containers to store the blends temporarily. All containers, including jars and bottles for creams, lotions, bath oils and salts can be purchased from retail stores or mail-order companies, or you can reuse previously used containers once they have been sterilized.

- GLASS MEASURING CUPS AND BOWLS: for measuring and mixing ingredients.

- GLASS RODS: for stirring. Using glass rods instead of wooden or plastic keeps the stirrer from absorbing the oils.

- LABELS: for every bottle or container, listing the ingredients, date of creation, and directions for use. Be creative in making the labels. For example, use rubber stamps, label-making computer software, or stickers. Some stores carry ready-made labels that are especially nice for hand-made products.

- MORTAR AND PESTLE: for mixing oils and coloring into solid materials.

Bathing Rituals

The oils can be used pure, using five to ten drops depending on the strength of the oil (for example, eucalyptus and peppermint would have a maximum of five drops, but up to ten drops could be used with milder lavender). If using the essences "neat" (without dilution), be sure to swirl them in the bath water before stepping in to avoid any possible skin irritations that might occur with direct skin contact. This method of using essential oils in the bath is good when you will be washing your hair and do not want the oily appearance that an oil blend would leave.

Since fragrance oils are not water soluble and do not fully disperse when added directly to bath water, diluting the oils is another effective way to use them. Fragrance oils may be diluted in unscented liquid soap for a bubble bath, combined with bath salts for a body scrub, or mixed with carrier oils as a bath oil. Usually 2 to 3 tablespoons of bath oil per bath is sufficient to receive the benefits of the oil. Also the addition of vitamin E to other oils acts as a preservative and prevents rancidness. It is excellent for all skin types and is reported to slow the aging process.

The combinations are numerous, but the results are the same: you should not do without luxurious bathing.

To make a bath oil that will totally disperse in the bath water, the carrier oil used must be turkey red oil or sulfated castor oil. It softens water without leaving a residue on your body or the tub, making it perfect to use in a Jacuzzi. Just substitute the turkey red oil for the carrier oil called for in any of the projects. Because this oil is not clear, it does not look as pretty in a clear bottle as sweet almond, for example. So use a colored bottle to disguise the color of the oil.

The temperature of the water can be used to complement the effect of the oils. Avoid hot baths because your skin will perspire and lose its ability to absorb the fragrance oils. Also, hot water dehydrates the skin and causes the oils to evaporate much more quickly. A warm or cool bath allows the oils to envelop your body delightfully when you slip into your bath and penetrate your skin, diffusing into the tissues. Breathe deeply, relax, and enjoy this time of quiet but necessary indulgence.

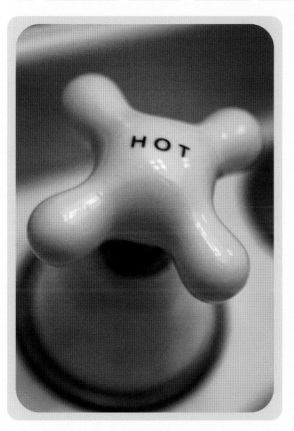

The No-Nonsense Woman's Nightcap

MAKES 2 TABLESPOONS

This is an addictive addition to your evening bath! For a restful sleep, use warm water to soothe the jagged edges.

MATERIALS
- 2 tablespoons carrier oil
- 15 drops lavender fragrance oil
- 5 drops chamomile fragrance oil
- 5 drops sweet orange fragrance oil

1. Blend the carrier oil with the fragrance oils.

2. Put the blend into the bath water just before getting in. (If added to the water while the tub is filling, much of the oils' essences will go up in the steam and very little will be left to be absorbed by the skin.) Swirl to disperse.

Morning Glory

MAKES 2 TABLESPOONS

Who needs coffee as a pick-me-up when you can start your day off right with this refreshing and motivating bath oil. For a refreshing, stimulating bath, use cool water.

MATERIALS

- 2 tablespoons carrier oil
- 12 drops lemon fragrance oil
- 4 drops rosemary fragrance oil
- 4 drops peppermint fragrance oil

1. Blend the carrier oil with the fragrance oils.

2. Put the blend into the bath water just before getting in. (If added to the water while the tub is filling, much of the oils' essences will go up in the steam and very little will be left to be absorbed by the skin.) Swirl to disperse.

Relaxing Bath for Gym Rats

MAKES 2 TABLESPOONS

This bath is perfect for relieving muscular aches and rheumatic pains. For a refreshing, relaxing bath, use warm water.

MATERIALS
- 2 tablespoons carrier oil
- 12 drops juniper fragrance oil
- 6 drops rosemary fragrance oil
- 4 drops lemon fragrance oil
- 4 drops eucalyptus fragrance oil

1. Blend the carrier oil with the fragrance oils.

2. Put the blend into the bath just before getting in. (If added to the water while the tub is filling, much of the oils' essences will go up in the steam and very little will be left to be absorbed by the skin.) Swirl to disperse.

Burned-Out Beauty Queen Bath Oil

MAKES 2 CUPS

High quality bath oils can be quite expensive to buy but can be made at a fraction of the cost. They are very easy to make but look extravagant. Bath oils are a great treatment for dry skin and especially healing in the winter months. They make great gifts for someone who has a very stressful life and likes to unwind with a long, luxurious bath.

MATERIALS

- 8 ounces sweet almond oil
- 8 ounces jojoba oil
- 20 drops lavender fragrance oil
- 10 drops geranium fragrance oil
- 8 vitamin E capsules
- dried lavender flower stems
- rose petals or buds
- decorative bottle
- cork top
- sealing wax
- raffia or ribbon

1. Blend the carrier oils, then the fragrance oils in a sterile container.

2. Break open the vitamin E capsules and empty them into the oil blend, stirring or shaking well.

3. Insert dried lavender stems and rose petals or buds into a clean and sterilized decorative bottle.

4. Using a funnel, fill bottle with the oil mixture.

5. Insert the cork top and seal the bottle with sealing wax.

6. Decorate the neck of the bottle with raffia or ribbon.

7. Label the bottle or provide a card that gives the ingredients and directions for use. Usually 2 to 3 tablespoons of bath oil per bath is sufficient to receive the benefits.

Recharging Bath Oil

MAKES 2 CUPS

This recipe can be used to fill a single bottle or several smaller ones. A separate container is always used for the blending before placing in sterilized, clean containers.

MATERIALS
- 8 ounces avocado oil
- 8 ounces grapeseed oil
- 20 drops of rosemary fragrance oil
- 10 drops of juniper fragrance oil
- 8 vitamin E capsules
- dried flowers
- decorative bottle
- cork top
- sealing wax
- raffia or ribbon

1. Blend the carrier oils, then the fragrance oils in a sterile container.

2. Break open the vitamin E capsules and empty them into the oil blend, stirring or shaking well.

3. Insert dried flowers into a clean and sterilized decorative bottle.

4. Using a funnel, fill bottle with the oil mixture.

5. Insert the cork top and seal the bottle with sealing wax.

6. Decorate the neck of the bottle with raffia or ribbon.

7. Label the bottle or provide a card that gives the ingredients and directions for use. Usually 2 to 3 tablespoons of bath oil per bath is sufficient to receive the benefits.

The Ski Bunny's Beauty Secret

MAKES 2 CUPS

Wish winter away with this soothing bath oil. The wheat germ oil will rehydrate skin dried out from the temperature shift between indoors and out.

MATERIALS

- 16 ounces wheat germ oil
- 20 drops sweet orange fragrance oil
- 10 drops geranium fragrance oil
- dried orange slices or rind
- decorative bottle
- cork top
- sealing wax
- raffia or ribbon

1. Blend the carrier oils with the fragrance oils in a sterile container.

2. Insert dried orange slices or rind into a clean and sterilized decorative bottle.

3. Using a funnel, fill bottle with the oil mixture.

4. Insert the cork top and seal the bottle with sealing wax.

5. Decorate the neck of the bottle with raffia or ribbon.

6. Label the bottle or provide a card that gives the ingredients and directions for use. Usually 2 to 3 tablespoons of bath oil per bath is sufficient to receive the benefits.

Eve's Sin-Free Bath

MAKES 2 CUPS

Bring the Garden of Eden to your bathtub. The fragrant floral blend of this bath oil will remind you of a garden in full bloom, while the water will soothe all your troubles away.

MATERIALS
- 8 ounces apricot kernel oil
- 8 ounces grapeseed oil
- 12 drops rosewood fragrance oil
- 8 drops ylang-ylang fragrance oil
- 8 drops lavender fragrance oil
- 2 drops patchouli fragrance oil
- 8 vitamin E capsules
- dried flowers
- decorative bottle
- cork top
- sealing wax
- raffia or ribbon

1. Blend the carrier oils, then the fragrance oils in a sterile container.

2. Break open the vitamin E capsules and empty them into the oil blend, stirring or shaking well.

3. Insert dried flowers into a clean and sterilized decorative bottle.

4. Using a funnel, fill bottle with the oil mixture.

5. Insert the cork top and seal the bottle with sealing wax.

6. Decorate the neck of the bottle with raffia or ribbon.

7. Label the bottle or provide a card that gives the ingredients and directions for use. Usually 2 to 3 tablespoons of bath oil per bath is sufficient to receive the benefits.

The Sweet Smell of Salt

MAKES 2 CUPS

Sea salt purifies your body and conditions the water and skin. With the addition of fragrance oils to the salt, you have a wonderful aromatic and therapeutic combination. Only a small amount of fragrance oils is needed to make a fragrant blend.

MATERIALS

- 1 cup baking soda
- 1 cup sea salt
- 5 drops jasmine fragrance oil (preferably essential and not synthetic)
- 3 drops rose absolute fragrance oil
- 5 drops ylang-ylang fragrance oil
- Few drops food coloring (optional)

1. Blend together baking soda, salt, and fragrance oils.

2. Add the food coloring separately to a small amount of the salts. Blend well then mix into remaining salts until desired color is visible.

3. Use immediately or store in a tightly sealed container. For each bath, use ¼ cup bath salts.

any moisture that might get into the container, place a small, muslin bag of rice in the bottom of the container. When properly sealed and stored, bath salts can last a long time.

MATERIALS

- 1 cup baking soda
- 1 cup sea salt
- 8 drops lavender fragrance oil
- 5 drops sweet orange fragrance oil
- Few drops blue food coloring

1. Blend together ½ cup baking soda, ½ cup salt, and food coloring.

2. Add fragrance oils to remaining ½ cup baking soda and ½ cup salt and blend well.

3. Add blue salts and mix just enough for all the salt to absorb the scent but not enough for the colors to blend thoroughly.

4. Use immediately or store in a tightly sealed container. For each bath, use ¼ cup bath salts.

Mermaid Bath Salts

MAKES 2 CUPS

Bath salts are easy to make, inexpensive, and make wonderful gifts. You can use decorative tins or jars with tight-fitting lids to hold the salts. Apothecary jars do nicely as containers for bath salts. To absorb

Bath Salts for Dancing Queens

MAKES 2 CUPS

Baking soda is also considered a salt and when added to the blend, provides another element for soothing aching muscles. It absorbs the fragrance oils easily and releases the scent when dissolved in bath water.

MATERIALS
- 1 cup baking soda
- 1 cup sea salt
- 4 drops eucalyptus fragrance oil
- 6 drops lavender fragrance oil
- 4 drops rosemary fragrance oil
- Few drops food coloring (optional)

1. Blend together baking soda, salt, and fragrance oils.

2. Add the food coloring separately to a small amount of the salts. Blend well then mix into remainder of salts until desired color is visible.

3. Use immediately or store in a tightly sealed container. For each bath, use ¼ cup bath salts.

Body Scrub for Cinderellas

MAKES 2 CUPS

Another way to use sea salt with essential oils is to blend them with a base oil to make an exhilarating exfoliant, which sloughs off dead cells on the surface of your skin, increases circulation, and gives the skin a healthy glow. Especially good carrier oils to use for this are sweet almond, apricot kernel, or grapeseed.

MATERIALS

- 1 ounce carrier oil
- 6 to 8 drops of fragrance oils of your choice
- 2 cups sea salt

1. To carrier oil, add fragrance oils and stir or shake to blend. Add to salt and mix well, making sure oil is thoroughly absorbed by the salt.

2. To apply, stand in bathtub or shower stall and massage the scrub gently over your body. Start with your feet and work your way up. Massage every part of the body except for the face and neck (too delicate); also go gently on breasts. Do not worry if a lot of the salt falls to the floor of the stall or tub; just pick up and keep using.

3. After you have finished the massage, shower or run a warm bath and wash the salt away. Your body will be invigorated and will have a lovely scent. After bathing, lavish your body with soothing, fragrant moisturizing lotion or cream.

Feisty Bath Fizzies

MAKES 1 FIZZY

A fun way to use bath salts in your bath is to make bath fizzies. These unique, sculpted salts are delightful as gifts and can be made in any number of shapes and sizes. The solid bath fizzy acts as a giant effervescent tablet in the bath, creating lots of bubbles, and releasing its fragrance as it dissolves. The bigger the molded fizzy, the longer it lasts. Try making an assortment of sizes, shapes, colors, and fragrances.

MATERIALS
- 1 cup baking soda
- ½ cup cornstarch

- ½ cup citric acid
- 8 drops lavender fragrance oil
- 4 drops lemon fragrance oil
- 3 drops Roman chamomile fragrance oil
- 10 drops food coloring (optional)
- soap mold

1. Mix baking soda, cornstarch, citric acid, and fragrance oils in a bowl.

2. Add food coloring to a small amount of the mixture in a separate bowl.

3. Add colored mixture to remaining mixture and blend.

4. Using a mister, spray the salt mixture with just enough water so that it holds together but does not start fizzing. Pack salt into a soap mold. Flip over onto a piece of waxed paper and allow molded fizzy to dry overnight.

Sexy-Scented Bath Vinegar

MAKES 8 CUPS

Apple cider vinegar is always a healthy addition to the bath. It is soothing, relieves dry, itchy skin, and restores natural pH to the skin. Combined with oils, this pleasantly scented vinegar provides the healthy benefits of apple cider vinegar plus the therapeutic properties of the fragrance oils. You can substitute the peppermint leaves and fragrance oils with rose, lavender, lemon verbena, or calendula petals and the corresponding fragrance oil. Try this recipe for yourself or bottle it up in a pretty bottle and give it as a gift.

MATERIALS
- 2 cups peppermint leaves
- 4 cups apple cider vinegar

- 4 cups distilled water
- 10 drops peppermint fragrance oil
- decorative glass bottle
- cork top
- sealing wax

1. Steep leaves in vinegar for 6 to 8 weeks. To shorten this time to 2 weeks, heat the vinegar before adding the herb.

2. After petals have steeped, strain the mixture. Try using a coffee filter in an old coffee-filter cone and place it on top of a container to drain.

3. To the strained liquid, add distilled water and fragrance oil. Cover.

4. Allow mixture to blend for a few days before pouring into a decorative glass bottle.

5. Insert the cork top and seal the bottle with sealing wax.

6. To use: As tub fills with water, pour 1 cup of vinegar into bath water. Or use as a compress by soaking a washcloth in the vinegar, then wringing it out, and applying it to the desired body area.

Belle of the Ball Bath Splash

MAKES ¾ CUP

After you have had a warm bath or shower, moisturize with one of the carrier oils, such as sweet almond oil or jojoba oil, while your body is still wet. The oil blends with the water left on your body and holds in the moisture. This is especially lovely in the evening when you can go to bed shortly after and let the oils work on making your skin soft and fragrant while you sleep.

If you are dressing to go out after your bath or shower, an after-bath splash lightly scents and refreshes the body.

MATERIALS
- ½ cup distilled water

- ¼ cup vodka
- 15 drops lavender fragrance oil
- 4 drops sandalwood fragrance oil
- 1 glass spritzer bottle
- Combine liquids and fragrance oils in spritzer bottle and shake until well mixed.

Soap Star

Be the star of your own soaps with these wonderful homemade creations that will tickle your senses and make washing yourself (or others!!!) a most enjoyable experience. Aromatic soaps are fabulous if you don't have the time to soak in a tub, but still need to perk yourself up. In other words, these soaps are perfect for the woman who wants to cut to the chase and get down to business. But no worries, not only do these soaps smell heavenly, they are pretty, too. Wrap them up and give them as a gift, or enjoy them for yourself! While soapmaking techniques can be quite elaborate, why add to your stress? This chapter tells you everything you need to know about making uncomplicated, yet festive soaps with the melt and pour technique. Then use your knowledge to be a bona fide soap star!

Soap Secrets

The ingredients and equipment needed to make soap are readily available at stores that sell crafts supplies, grocery stores, health food stores, and drug stores, and can be ordered from catalogs and the Internet. Many items may already be in your kitchen.

MELT-AND-POUR SOAP BASES

There are many types of melt-and-pour soap bases available. The basic three are clear glycerin soap base, whitened glycerin soap base, and white coconut oil soap base. Soap base variations are the result of additives mixed into the bases, such as olive oil, coconut oil, hemp oil, and colorants. All melt-and-pour bases are designed to melt easily in the microwave or a double boiler and to be poured into molds.

People who are familiar with hand-milled and cold-processed soapmaking techniques will want to note the differences between those techniques and the melt-and-pour technique. First, water is never added to melt-and-pour soap bases; it makes them slimy and prevents the soap from hardening properly. Also, melt-and-pour soaps set up immediately with little curing time, enabling you to create handmade

soaps instantly. Moreover, the creative and design possibilities are far greater with the melt-and-pour soap method.

CLEAR GLYCERIN SOAP BASE

High quality glycerin soap should be gentle enough for all skin types and have little scent (so it is ready for your own fragrance blends). If you wish a cloudy glycerin soap to be even more transparent when molded, simply melt, let harden, and remelt before pouring into your mold. The second melting helps remove excess moisture from the soap. Most glycerin soaps have a melting point between 135°F and 145°F.

WHITENED GLYCERIN SOAP BASE

Most white melt-and-pour soap bases are glycerin soap whitened with titanium dioxide, a white mineral pigment. They are sometimes called "coconut oil soap" when extra coconut oil is added. Whitened glycerin soap has a milky translucent look when melted and a lower melting point than clear glycerin soap. You can make your own whitened glycerin soap by adding white colorant to a clear glycerin soap base. Using the white colorant enables you to control the effects; you can add a little colorant for a translucent bar or more colorant for an opaque bar.

WHITE COCONUT OIL SOAP BASE

This base is true coconut oil soap, not whitened glycerin soap with added coconut oil. Made with coconut and vegetable oils, this pH-balanced soap is enriched with vitamin E and makes your skin feel soft, clean, and healthy. The coconut oil helps create a protective barrier that keeps the skin supple. It is easy to know if you are working with true coconut oil soap: When melted, it becomes clear; when fully hardened, it is a bright white soap. Coconut oil soap has a high melting point (190°F), which makes for a wonderful hard-finished soap. This high temperature can melt some molds; make sure you cool the melted soap by stirring before pouring it into a mold.

OTHER GLYCERIN SOAP BASES

Most other soap bases are glycerin-based. Added ingredients provide different qualities. Some examples include:

- **Glycerin with coconut oil:** This soap base is an opaque white and may have a light coconut scent.

- **Glycerin with avocado and cucumber:** This opaque, pale-green soap base has been enriched with vitamin E. A suspension formula keeps the additives uniformly distributed within the soap bar.

- **Glycerin with olive oil:** This soap base is a translucent natural amber or green color. The added olive oil hydrates the skin. It contains vitamin E and antioxidants and may feature a suspension formula.

- **Glycerin with hemp oil:** Silky and soothing, this translucent natural green-colored soap base is good for dry skin and recognized for its ability to repair damaged skin cells.

- **Glycerin with goat's milk:** This opaque white soap base offers gentle moisturizing while maintaining the skin's natural pH level.

- **Colored glycerin soap bases:** These include clear colored soap bases, opaque colored soap bases, and bright florescent colored soap bases.

- **Frosting soap base:** A specialty soap base formulated to accent finished soaps, it has the look of fluffy white confectioner's icing.

SOAP BASE FORMS

Soap bases come in different forms, including blocks and bars, soap sheets, preformed inserts (butterflies, stars, hearts, etc.), soap cubes, soap curls, soap noodles, and soap shavings.

CHOOSING QUALITY MELT-AND-POUR SOAP BASES

For best results and a high quality finished soap bar, purchase soap bases that are made with vegetable oils and without animal by-products or fillers such as wax and alcohol. The cheaper the soap, the more likely it is to contain wax and fillers. Wax and fillers cost less than other soap ingredients and account for the wide range of prices for melt-and-pour soap bases. Adding wax and fillers to glycerin soap base greatly diminishes the soap's cleansing qualities and lathering capabilities, and it can affect the smell and look of your finished soap.

Some melt-and-pour glycerin soap bases contain added alcohol as a means of removing moisture, thus making the soap more transparent. However, added alcohol creates a strong-smelling soap with several safety and health concerns. Alcohol is very flammable and can ignite when the soap base is melted. Alcohol is also considered a skin irritant; repeated use will dry out your skin. Soap made with alcohol should not be used by anyone who has sensitive skin or dry skin.

- Always use recommended safe ingredients. Just because an ingredient is natural, it does not mean it is safe to use in soap.

- Too much of an additive may soften your soap or make it scratchy and uncomfortable to use. Follow the recipe and use the recommended amount.

- Do not use fresh vegetables or fruits in melt-and-pour soap bases. Any material that is not properly dried or preserved can cause your soap to become rancid very quickly.

- Beware of soap recipes that include additives that make the soap look good but seem impractical or do not impart added benefits. For example, potpourri may look good sprinkled in a clear soap, but the large petals will clog your drain and scratch your skin. Additionally, potpourri may contain ingredients that would be harmful when in contact with your skin.

- When adding extra oils, add a natural preservative for a longer lasting bar. Natural preservatives include citric acid, vitamin E oil, and grapefruit seed extract.

ADDITIVES FOR SOAPS

Additives such as oatmeal, dried herbs, and oils can be added to soap bases to nourish, soften, and add gentle scrubbing properties. Each ingredient offers its own unique characteristics to your products. Many additives also can add interesting colors and textures to your soap.

Remember these points when including additives in your soaps:

A FEW CAUTIONS ABOUT ADDITIVES

Just because plants are natural does not mean they are safe. Many of the world's deadliest poisons come from plants. Herbs must be used with caution, as many are potentially harmful irritants or can cause allergic reactions. Almost every additive, natural or synthetic, can trigger someone's allergy or irritate someone's sensitive skin.

Although these reactions are annoying, it is possible to avoid a recurrence by eliminating the offending ingredient. You can perform a simple skin test to make sure you are not allergic to a soap by rubbing a small amount of the additive on the tender area on the inside of your elbow. If you are sensitive to any ingredient, your skin will develop redness or a slight rash.

- Avoid using soaps with abrasive fillers on your face. Save those soaps for rough spots such as elbows, knees, and hands. Soaps made from melt-and-pour soap bases that include fragrance oils and extra fats for moisturizing are best for cleansing your face.

- For natural additives, use those suggested on the list of additives. If you would like to explore this area further, there are many books available to educate you in the safe use of natural products.

- Use only herbs and flowers that are clean and free of insecticides and chemicals. Spray residues on plant material can irritate your skin. If plants that you have grown yourself are not available, purchase botanicals (ideally organically grown) from a natural food store or buy them fresh from the market and dry them yourself. Do not use dried botanicals sold for potpourri or dried flower arranging in soap; these botanicals are not required to be food safe and may contain harmful dyes or chemicals.

- Cocoa butter, coconut oil, and almonds may produce a reaction in those allergic to chocolate and nuts.

- Natural emollients such as lanolin and glycerin may cause a reaction in those with sensitive skin.

- Honey, bee pollen, and beeswax may cause a reaction in those allergic to pollen. Do not use unpasteurized honey to scent products that will be used on infants.

SAFE ADDITIVES

BOTANICALS (dried)

- Bergamot leaf and flower
- Calendula petal
- Chamomile blossoms
- Coffee beans
- Cranberries
- Crushed apricots
- Eucalyptus leaves
- Ginger
- Green tea
- Kelp
- Lavender buds
- Lemon peel
- Lemon verbena leaves
- Lemongrass
- Loofah
- Orange peel
- Peppermint leaves
- Pine needles
- Poppy seeds
- Rose hips
- Rose petals
- Rosemary leaves
- Sage leaves
- Sandalwood
- Strawberry leaves
- Witch hazel (leaves and liquid extract)

NUTS

- Almonds
- Coconuts
- Hazelnuts

OILS

- Sweet almond oil
- Cocoa butter
- Coconut oil
- Olive oil
- Palm oil
- Shea nut butter
- Spices
- Allspice
- Anise
- Cardamom
- Cinnamon
- Cloves
- Paprika
- Tumeric
- Vanilla pods

MISCELLANEOUS

- Aloe vera gel
- Bee pollen
- Cornmeal
- French clay
- Glycerin (liquid)
- Honey
- Milk (powdered whole cow's milk, goat's milk, and buttermilk)
- Oatmeal
- Oyster shell
- Pumice
- Rosewater
- Salt (fine sea, rock)
- Tapioca
- Wheat bran

PRESERVATIVES

- Citric acid
- Grapefruit seed extract
- Vitamin E (break open a capsule and squirt oil into melted soap)

MIXING ADDITIVES

Measure your additives and blend them together before adding them to the melted soap. Because additives such as powdered milk or spices can clump up, it is best to mix them with liquid glycerin before adding them to the melted soap base to help disperse them evenly. Additives such as powdered spices, seeds, and grains will either sink to the bottom of the mold or float on the surface of the soap, an effect that can give the soap a natural, whimsical look. To achieve this effect, you can add the additives to the melted soap before pouring or place them in the bottom of the mold and pour the melted soap base on top.

To create soap with additives suspended throughout, there are two options. You can buy soap bases that have been especially formulated to suspend additives. If you use one of those, just add the additives, mix, and pour into the mold. Or, after adding the additives to the melted soap base, gently stir the soap with a spoon to slowly cool and thicken the base. As soon as the soap starts to thicken, pour it into the mold. The soap will harden with the additives suspended throughout the soap. Be careful not to let the soap thicken so much that it will not pour. If this happens, remelt and start again. When mixing a large amount of soap, it is helpful to grate some of the same type of soap base and have a pile of it on hand to add to the soap; this will cool and quickly thicken the melted soap.

COLORANT

Color is an important ingredient of soap's allure. You can add spices and dried herbs for a natural soap or cosmetic grade colorants for a brightly colored bar.

Soaps colored with natural powders have a wholesome, country look with attractive warm brown and tan hues. Natural powders include cocoa powder, dried herbs, and ground spices.

Cosmetic-grade colorants are available as both solids and liquids, and they can be found in the fragrance crafting departments of craft stores or from sellers of soapmaking supplies. High-quality

cosmetic-grade colorants create true, clean colors and are excellent for blending and creating many different hues. The liquid colors come in red, blue, yellow, orange, green, white, and black. Solid colorants also are available in a wide range of hues.

Food coloring is not suitable for soapmaking because the color quickly fades.

For a sparkling effect, try cosmetic-grade glitters.

COLOR MIXING BASICS

The best way to learn about color mixing is through personal experience and experimentation. To get started, it helps to understand a few basic principles.

Primary colors are red, blue, and yellow. Secondary colors are mixes of primary colors. Green is made from yellow and blue; purple from red and blue, and orange from red and yellow. Intermediate colors are mixes of a primary color with a neighboring secondary color on a color wheel. For example, lime green is a mixture of yellow and green.

Complementary colors are colors that are opposite one another on the color wheel; red is the complement of green, purple is the complement of yellow, blue is the complement of orange. When you mix a color with its complement, the result is a dulling or muting of the color, making it less intense. Dusty plum, for example, is purple plus a touch of its complement, yellow; golden ocher is yellow plus a touch of its complement, purple.

Shades are made by adding black to a color. Tints are made by adding white.

COLORANT TIPS
Points to consider when adding colorants

- For the soap recipes in this book, colorant amounts are listed as drops of liquid colorants made for soap.
- Because the results can vary a great deal depending on the size or number of drops you use, take notes while mixing colors and keep a record of your results. Also note the type of soap base you use. In a clear glycerin soap base, colors appear clear and jewel-toned, but in a white coconut oil soap base the same colorants create soft pastels.
- Some colorants are transparent and leave a glycerin soap base clear, while others add an opaque tint to your soap. Test your colorants before making a large batch to avoid disappointment.
- Some fragrance oils have a strong color and will tint your soap. For example, if you want to make a clear blue soap, adding vanilla fragrance oil (which has a strong amber hue) to a clear glycerin soap base will give the soap a greenish tint.
- Adding too much colorant makes the colors blend faster into one another when making two-color soaps.

EQUIPMENT & TOOLS

The basic equipment and tools for creating melt-and-pour soaps are standard kitchen items that you probably already own.

- **Electric spice/coffee grinder:** A small electric grinder is useful when grinding small amounts of additives, such as spices, nuts, or oatmeal. Clean the grinder after each use by grinding a piece of fresh bread or some rice (the bread or rice soaks up oils), then wipe with a paper towel.

- **Glass measuring cups:** Use heat-resistant glass measuring cups to melt soap bases in the microwave or on the stovetop (with a pan to make a double boiler). You need 1-, 2-, and 4-cup sizes.

- **Glass droppers:** You need three to five glass droppers for measuring fragrance oils. Do not use plastic droppers. They cannot be cleaned completely (so you could contaminate your oils with other scents), and some essential oils will dissolve the plastic.

- **Kitchen tools:** Many kitchen tools come in handy for special effects. A melon baller can be used to scoop out uniform holes from soap; a crinkle cutter can be used to slice soaps with a decorative wavy cut.

- **Knife:** Use a sharp knife to cut blocks of melt-and-pour soap base into smaller pieces and to slice your finished molded soaps. Have a variety of knives handy, from small paring knives to large butcher knives.

- **Measuring spoons:** You need a set of metal or plastic measuring spoons to measure additives.

- **Mixing spoons:** Metal or wooden kitchen spoons are needed to mix melted soap. Since metal spoons will not transfer fragrances, they are safe for food use after cleaning. If you use wooden spoons, clearly mark them "For fragrance crafting only" and do not use them again for cooking because the wood will retain scents and transfer them to your food.

- **Mortar and pestle:** A mortar and pestle can be preferable for grinding because you have more control. A mortar is easier to clean, too; just wipe it out with a paper towel after use.

- **Paper cups:** Use small paper cups to hold pre-measured additives and to use as equal-size risers for tray molds that will not sit level on a flat surface.

- **Saucepan** (for the stovetop method): Use a large metal pan to make a simple double boiler by filling the saucepan with water and placing a large heat-resistant glass measuring cup in it to melt the soap bases.

- **Soap beveller:** This is really a cheese plane that can be used to bevel soap edges. It can also be used to clean soap surfaces and make soap curls.

- **Waxed paper:** Use waxed paper to protect your work area when pouring and molding soaps.

CARING FOR TOOLS AND EQUIPMENT

Clean glass and metal tools thoroughly after use. After cleaning, they can be used for food preparation. Plastic and wood items used for soapmaking should not be used for food.

Soap Molds

Molds help you make your melt-and-pour soaps appear more professional and fancy. It is a good idea to have a selection of soap molds that includes traditional soap shapes and fun theme shapes.

Most molds can tolerate temperatures of 135°F to 145°F. Overheated soap sometimes warps even the best soap molds. This can happen especially with coconut oil soap, which has a high melting point (190°F). To avoid warping, cool the soap by stirring or set the molds in a shallow cold-water bath for high temperature pours.

Mold designs show up clearer and crisper in hard soap than in soft soap. Adding palm oil or cocoa butter can harden soft glycerin soaps.

CAUTION: Be careful when choosing plastic containers for molding your soaps. Some cannot take the high temperature of the melted soap and can melt and collapse, causing a spill.

MOLDS ESPECIALLY FOR SOAPMAKING

High-heat plastic soap molds are, overall, the best and safest for soapmaking. Their deep, clean, smooth contours allow you to create professional-looking soaps safely and unmold them easily. The molds come in individual shapes or in trays of fancy motifs. They are designed to last through repeated moldings and will withstand the high temperatures of melt-and-pour soap bases without warping or melting. Good-quality soap molds do not require pretreatment to release the hardened soaps and are self-leveling. Large loaf-style molds are available in a variety of sizes and shapes. Look for very small shapes designed especially for chunk-style soaps at crafts stores.

Tube molds are available in plastic and metal, in both tubular and two-part, snap-together styles. Having a selection of large and small tube molds in basic shapes enables you to make a wide variety of designs.

Individual resin casting molds are an excellent size and depth for single bars of homemade soap.

The number of ounces of melted soap the mold holds is printed into the bottom of each mold.

CANDLE MOLDS

Some plastic candle molds are suitable for soapmaking. Since candle molds are designed to withstand the high temperatures of melted wax, they can withstand hot soap without melting. It can be difficult, however, to unmold the soap. For best results, choose low, wide molds rather than long, skinny ones and always use a mold release.

Metal candle molds can handle the high temperatures, but most soap chemically reacts with metal, especially aluminum. This reaction causes the metal to corrode, and the corrosion discolors the soap and will eventually destroy the metal molds. Because they are rigid, metal molds need a mold release.

LATEX MOLDS

Rubber latex molds make beautiful three-dimensional soaps. (Rubber molds for candlemaking also can be used.) Because flexible red rubber molds can transfer the color to the soap, choose rubber molds.

You can also make your own rubber molds using a liquid rubber mold builder that takes an impression of a variety of objects. Imagine using a shell, glass figurine, or stone carving to make your own signature soap mold!

PLASTIC FOOD-STORAGE CONTAINERS

Small sandwich or storage containers (4" by 6") will accommodate loaf-style soaps. Try to find ones with no design on the inside bottom and with nicely rounded corners. Caution: Some plastic containers cannot withstand the high temperature of melted soap base and will melt and collapse, spilling the hot soap. Look for containers that are dishwasher and microwave safe; be wary of disposable plastic containers and takeout food containers, which are intended for one-time use and are not as sturdy. When using plastic containers, always stir the soap before pouring, to cool it.

CANDY AND PLASTER MOLDS

Plaster and candy molds can warp, as they cannot take the high temperatures of the melt-and-pour soap. If hot melted soap is poured into them, they melt or deform, spilling their contents, which can result in serious burns if one is not careful. Many plastic molds designed to keep packaged food from crushing are tempting because of the wonderful designs and deep shapes, but the plastic will warp and melt when the soap is added. For this reason, these molds are unsuitable for soapmaking.

when cold as a soft paste. Apply a thin, even layer, or it can mar the smoothness of the finished soap.

Vegetable oil in liquid and spray forms is also used as a mold release. However, even with a thin film on the mold, it tends to make the soap a bit greasy and the oil can turn rancid over time, altering the fragrance of the soap.

MOLD RELEASES

To release soap from the mold, coat the inside of the mold with a thin layer of petroleum jelly before pouring in the soap. Widely used in cosmetics as an emollient and barrier cream, petroleum jelly does not leave a sticky residue on the soap.

Another suggestion for a soap release comes from soap mold manufacturers themselves: use one part (by weight) paraffin wax and stir in three parts baby oil. This is best used when hot, but it can be applied

UNMOLDING TIPS

- **Use gentle pressure.** release the soap from tray molds by using gentle pressure from your thumbs on the back of the mold. You can easily damage molds with improper handling.

- **Leave it in the mold.** If the soap is allowed to remain in the mold for twelve to twenty-four hours after cooling down, it will release much more easily than if unmolded immediately upon cooling.

- **Try the freezer.** If you still find it difficult to release the soap from a mold, try placing it in the freezer for ten minutes, remove, and try again.

PREPARING TO MAKE SOAP

To calculate the amount of soap needed to fill your mold, fill the mold with water, then pour the water into a measuring cup to measure.

Always melt at least one additional ounce of soap to account for the soap that will cling to the inside of the measuring cup.

Slice the soap base into small pieces for quick, easy melting.

Make sure all the bowls, measuring cups, and mixing spoons are completely dry.

Never add water to melt-and-pour soap bases.

SOAP MELTING TIPS

- It is harmless to remelt the soap. If you melt more soap base than is needed to fill your chosen mold, pour the extra soap base into a spare mold or plastic container, let it harden, release, and remelt when making another project. (It is a good idea to always have an extra mold on hand in case this happens.)

- Repeated remelting of clear glycerin soap base makes it more transparent, as the excess moisture evaporates. Repeated remelting of coconut oil soap base makes the resulting soap harder.

- When cleaning up, do not put your measuring cups or spoons in the dishwasher. Soap base is designed to make lots of luxurious bubbles, and the soap left on the equipment could foam and cause the dishwasher to leak. Roll up your sleeves and wash the few pieces of equipment by hand.

- Experimenting by trial and error helps you understand the process.

Hooray for Bollywood Soap!

MAKES 2 BARS

Take a trip to the tropics with this exotic blend of scents.

MATERIALS
- 3 ounces whitened glycerin soap base
- 1 teaspoon powdered milk powder
- 1 teaspoon coconut oil
- 5 drops mango fragrance oil
- 5 drops coconut fragrance oil
- 2 round molds, 2½ ounces each

1. Place the soap base into a heat-resistant glass measuring cup and microwave for approximately 10 to 25 seconds. If you melt the soap in a double boiler on the stove, adjust the heat to keep the soap at a constant liquid point. Do not let the soap heat for more than 10 minutes.

2. Remove from the microwave or stovetop and stir lightly to melt completely any remaining soap pieces. Do not leave the mixing spoon in the soap while heating in the microwave or when melting on the stovetop.

3. Immediately add the powdered milk and coconut oil to the melted soap. Stir gently to mix. If the soap starts to solidify, reheat it to remelt it.

4. Add the fragrance oils one drop at a time until the desired level of scent is achieved.

5. Pour the soap into the molds immediately.

6. Let the soap cool and harden completely before removing from the mold. The soap will pop out easily when completely set. For a fast set, allow the soap to set in the refrigerator until cooled.

The Ingenue's Secret of Youth

MAKES 2 BARS

Olive oil does wonders for the skin. Revitalize yourself and show off a radiant glow with this soap.

MATERIALS
- 6 ounces clear glycerin soap base with added olive oil
- 10 drops ylang-ylang fragrance oil
- 10 drops rose fragrance oil
- 2 bee-and-honeycomb molds, 3 ounces each

1. Place the soap base into a heat-resistant glass measuring cup and microwave for approximately 30 to 45 seconds. If you melt the soap in a double boiler on the stove, adjust the heat to keep the soap at a constant liquid point. Do not let the soap heat for more than 10 minutes.

2. Remove from the microwave or stovetop and stir lightly to melt completely any remaining soap pieces. Do not leave the mixing spoon in the soap while heating in the microwave or when melting on the stovetop.

3. Add the fragrance oils one drop at a time until the desired level of scent is achieved.

4. Pour the soap into the molds immediately.

5. Let the soap cool and harden completely before removing from the mold. The soap will pop out easily when completely set. For a fast set, allow the soap to set in the refrigerator until cooled.

You Go Girl Soap

MAKES 2 BARS

Increase your energy with this feisty, fun blend of scents and shapes.

MATERIALS

- 1 ounce clear glycerin soap base
- 3 ounces whitened glycerin soap base
- 3 drops green colorant
- 3 drops yellow colorant
- 10 drops rosemary fragrance oil
- 5 drops peppermint fragrance oil
- 2 rectangular molds, 2 ounces each, with embossed frog and lizard

1. Place the clear soap base into a heat-resistant glass measuring cup and microwave for approximately 10 seconds. If you melt the soap in a double boiler on the stove, adjust the heat to keep the soap at a constant liquid point. Do not let the soap heat for more than 10 minutes.

2. Remove from the microwave or stovetop and stir lightly to melt completely any remaining soap pieces. Do not leave the mixing spoon in the soap while heating in the microwave or when melting on the stovetop.

3. Immediately add 2 drops of the green colorant to the melted soap. Stir gently to mix. If the soap starts to solidify, reheat it to remelt it.

4. Pour the soap into the embossed areas of the molds immediately.

5. Let it harden a bit, then take a piece of hard plastic (an old credit card works great) and scrape away any soap that spilled over and out of the embossed design area.

6. Place the whitened soap base into a heat-resistant glass measuring cup and microwave for approximately 10 to 25 seconds. If you melt the soap in a double boiler on the stove, adjust the heat to keep the soap at a constant liquid point. Do not let the soap heat for more than 10 minutes.

7. Remove from the microwave or stovetop and stir lightly to melt completely any remaining soap pieces. Do not leave the mixing spoon in the soap while heating in the microwave or when melting on the stovetop.

8. Immediately add 1 drop of the green colorant and the yellow colorant to the melted soap. Stir gently to mix. If the soap starts to solidify, reheat it to remelt it.

9. Add the fragrance oils one drop at a time until the desired level of scent is achieved.

10. Pour the soap into the molds immediately.

11. Let the soap cool and harden completely before removing from the mold. The soap will pop out easily when completely set. For a fast set, allow the soap to set in the refrigerator until cooled.

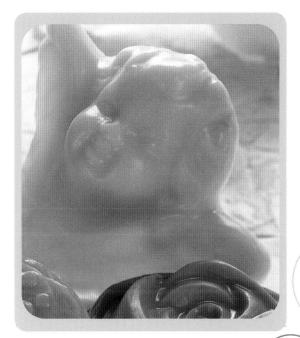

Lighten-Up Bar

MAKES 1 BAR
Boost your spirits with this soft, sweet bar of sudsy goodness.

MATERIALS

- 3 ounces glycerin soap base with added goat's milk
- 10 drops vanilla fragrance oil
- 6 drops sweet orange fragrance oil
- 1 angel mold, 3 ounces

1. Place the soap base into a heat-resistant glass measuring cup and microwave for approximately 10 to 25 seconds. If you melt the soap in a double boiler on the stove, adjust the heat to keep the soap at a constant liquid point. Do not let the soap heat for more than 10 minutes.

2. Remove from the microwave or stovetop and stir lightly to melt completely any remaining soap pieces. Do not leave the mixing spoon in the soap while heating in the microwave or when melting on the stovetop.

3. Add the fragrance oils one drop at a time until the desired level of scent is achieved.

4. Pour the soap into the mold immediately.

5. Let the soap cool and harden completely before removing from the mold. The soap will pop out easily when completely set. For a fast set, allow the soap to set in the refrigerator until cooled.

Naughty Girl Soap

MAKES 2 GUEST-SIZE BARS

This brilliant blend of scents is wonderfully whimsical.

MATERIALS
- 4 ounces whitened glycerin soap base
- 2 drops orange colorant
- 10 drops vanilla fragrance oil
- 10 drops watermelon fragrance oil
- 5 drops rose fragrance oil
- 2 round molds, 1½ ounces each, with embossed fish

1. Place the soap base into a heat-resistant glass measuring cup and microwave for approximately 20 to 30 seconds. If you melt the soap in a double boiler on the stove, adjust the heat to keep the

soap at a constant liquid point. Do not let the soap heat for more than 10 minutes.

2. Remove from the microwave or stovetop and stir lightly to melt completely any remaining soap pieces. Do not leave the mixing spoon in the soap while heating in the microwave or when melting on the stovetop.

3. Pour 1 ounce of the soap into the embossed areas of the molds immediately.

4. Let it hardened a bit, then take a piece of hard plastic (an old credit card works great) and scrape away any soap that spilled over and out of the embossed design area.

5. Add the colorant to the remaining melted soap. Stir gently to mix. If the soap starts to solidify, reheat it to remelt it.

6. Add the fragrance oils one drop at a time until the desired level of scent is achieved.

7. Pour the soap into the molds immediately.

8. Let the soap cool and harden completely before removing from the mold. The soap will pop out easily when completely set. For a fast set, allow the soap to set in the refrigerator until cooled.

Peaches & Cream Complexion Bar

MAKES 2 BARS

The vitamin E and aloe vera gel in this fantastic soap will soothe the skin.

MATERIALS

- 8 ounces whitened glycerin soap base
- 2 capsules vitamin E
- 1 teaspoon aloe vera gel
- 4 drops orange colorant
- 2 drops red colorant
- 15 drops aloe vera fragrance oil
- 10 drops rose fragrance oil
- 2 rose-shaped molds, 3¾-ounce

1. Place the soap base into a heat-resistant glass measuring cup and microwave for approximately one minute. If you melt the soap in a double boiler on the stove, adjust the heat to keep the soap at a constant liquid point. Do not let the soap heat for more than 10 minutes.

2. Remove from the microwave or stovetop and stir lightly to melt completely any remaining soap pieces. Do not leave the mixing spoon in the soap while heating in the microwave or when melting on the stovetop.

3. Immediately add the vitamin E, the aloe vera gel, and the colorants to the melted soap. Stir gently to mix. If the soap starts to solidify, reheat it to remelt it.

4. Add the fragrance oils one drop at a time until the desired level of scent is achieved.

5. Pour the soap into the molds immediately.

6. Let the soap cool and harden completely before removing from the mold. The soap will pop out easily when completely set. For a fast set, allow the soap to set in the refrigerator until cooled.

Cupid's Cure-All

MAKES 1 BAR

A wonderful way to say "I love you," this soap is bound to attract even the most resistant sweethearts!

MATERIALS

- 4 ounces clear glycerin soap base
- 4 drops red colorant
- 1 drop black colorant
- 10 drops rose fragrance oil
- 5 drops clove fragrance oil
- 5 drops ylang-ylang fragrance oil
- 1 embossed heart mold, 4 ounces

1. Place the soap base into a heat-resistant glass measuring cup and microwave for approximately 20 to 30 seconds. If you melt the soap in a double boiler on the stove, adjust the heat to keep the soap at a constant liquid point. Do not let the soap heat for more than 10 minutes.

2. Remove from the microwave or stovetop and stir lightly to melt completely any remaining soap pieces. Do not leave the mixing spoon in the soap while heating in the microwave or when melting on the stovetop.

3. Immediately add the colorants to the melted soap. Stir gently to mix. If the soap starts to solidify, reheat it to remelt it.

4. Add the fragrance oils one drop at a time until the desired level of scent is achieved.

5. Pour the soap into the mold immediately.

6. Let the soap cool and harden completely before removing from the mold. The soap will pop out easily when completely set. For a fast set, allow the soap to set in the refrigerator until cooled.

Calorie-Free Comfort Bar

MAKES 1 BAR

The mixture of fragrances creates a scent that will calm you and help relieve stress.

MATERIALS
- 3 ounces clear glycerin soap base
- 3 drops orange colorant
- 6 drops vanilla fragrance oil
- 5 drops jasmine fragrance oil
- 1 butterfly mold, 3 ounces

1. Place the soap base into a heat-resistant glass measuring cup and microwave for approximately 10 to 25 seconds. If you melt the soap in a double boiler on the stove, adjust the heat to keep the soap at a constant liquid point. Do not let the soap heat for more than 10 minutes.

2. Remove from the microwave or stovetop and stir lightly to melt completely any remaining soap pieces. Do not leave the mixing spoon in the soap while heating in the microwave or when melting on the stovetop.

3. Immediately add the colorant to the melted soap. Stir gently to mix. If the soap starts to solidify, reheat it to remelt it.

4. Add the fragrance oils one drop at a time until the desired level of scent is achieved.

5. Pour the soap into the mold immediately.

6. Let the soap cool and harden completely before removing from the mold. The soap will pop out easily when completely set. For a fast set, allow the soap to set in the refrigerator until cooled.

Serenity Now

MAKES 1 BAR

Who has patience, especially when upset? Calm yourself at once with this fragrant blend.

MATERIALS
- 4¼ ounces clear glycerin soap base
- 4 drops green colorant
- 10 drops sweet orange fragrance oil
- 5 drops patchouli fragrance oil
- 1 arabesque mold, 4¼ ounces

1. Place the soap base into a heat-resistant glass measuring cup and microwave for approximately 20 to 35 seconds. If you melt the soap in a double boiler on the stove, adjust the heat to keep the soap at a constant liquid point. Do not let the soap heat for more than 10 minutes.

2. Remove from the microwave or stovetop and stir lightly to melt completely any remaining soap pieces. Do not leave the mixing spoon in the soap while heating in the microwave or when melting on the stovetop.

3. Immediately add the colorant to the melted soap. Stir gently to mix. If the soap starts to solidify, reheat it to remelt it.

4. Add the fragrance oils one drop at a time until the desired level of scent is achieved.

5. Pour the soap into the mold immediately.

6. Let the soap cool and harden completely before removing from the mold. The soap will pop out easily when completely set. For a fast set, allow the soap to set in the refrigerator until cooled.

Sinfully Delicious Massage Bar

MAKES 1 BAR

Apricot is a fabulous scent without being too fruity or overwhelming.

MATERIALS

- 3¾-ounce ounces glycerin soap base with added olive oil and suspension formula
- 1 teaspoon ground apricot seed
- 2 drops chamomile fragrance oil
- 1 rectangular massage mold, 3¾ ounces

1. Place the soap base into a heat-resistant glass measuring cup and microwave for approximately 15 to 25 seconds. If you melt the soap in a double broiler on the stove, adjust the heat to keep the soap at a constant liquid point. Do not let the soap heat for more than 10 minutes.

2. Remove from the microwave or stovetop and stir lightly to melt completely any remaining soap pieces. Do not leave the mixing spoon in the soap while heating in the microwave or when melting on the stovetop.

3. Immediately add the apricot seed to the melted soap. Stir gently to mix. If the soap starts to solidify, reheat it to remelt it.

4. Add the fragrance oil one drop at a time until the desired level of scent is achieved.

5. Pour the soap into the mold immediately.

6. Let the soap cool and harden completely before removing from the mold. The soap will pop out easily when completely set. For a fast set, allow the soap to set in the refrigerator until cooled.

Work It! Soap

MAKES 1 BAR

Love the gym, but hate the locker-room smell? Your troubles are over with this combination, perfect for a post-workout cleansing.

MATERIALS

- 1 ounce whitened glycerin soap base
- 3 ounces clear glycerin soap base with added hemp oil
- 1 teaspoon dried peppermint leaves
- 5 drops earth fragrance oil
- 3 drops pine fragrance oil
- 3 drops spearmint fragrance oil
- 1 rectangular massage mold, 4 ounces

1. Place the whitened soap base into a heat-resistant glass measuring cup and microwave for approximately 10 seconds. If you melt the soap in a double boiler on the stove, adjust the heat to keep the soap at a constant liquid point. Do not let the soap heat for more than 10 minutes.

2. Remove from the microwave or stovetop and stir lightly to melt completely any remaining soap pieces. Do not leave the mixing spoon in the soap while heating in the microwave or when melting on the stovetop.

3. Pour the soap into the embossed areas of the mold immediately.

4. Let it harden a bit, then take a piece of hard plastic (an old credit card works great) and scrape away any soap that spilled over and out of the embossed design area.

5. Place the clear glycerin soap base with added hemp oil into a heat-resistant glass measuring cup and microwave for approximately 10 to 25 seconds. If you melt the soap in a double boiler on the stove, adjust the heat to keep the soap at a constant liquid point. Do not let the soap heat for more than 10 minutes.

6. Remove from the microwave or stovetop and stir lightly to melt completely any remaining soap pieces. Do not leave the mixing spoon in the soap while heating in the microwave or when melting on the stovetop.

7. Immediately add the dried peppermint leaves to the melted clear glycerin soap base with added hemp oil. Stir gently to mix. If the soap starts to solidify, reheat it to remelt it.

8. Add the fragrance oils one drop at a time until the desired level of scent is achieved.

9. Pour the soap into the mold immediately.

10. Let the soap cool and harden completely before removing from the mold. The soap will pop out easily when completely set. For a fast set, allow the soap to set in the refrigerator until cooled.

Make My Day! Massage Bar

- -

MAKES 1 BAR

This soap cleans and relaxes you at the same time. Who could ask for anything more?

MATERIALS

- 3 ounces glycerin soap base with added hemp oil
- 5 drops eucalyptus fragrance oil
- 5 drops peppermint fragrance oil
- 1 round massage mold, 3 ounces

1. Place the soap base into a heat-resistant glass measuring cup and microwave for approximately 10 to 25 seconds. If you melt the soap in a double boiler on the stove, adjust the heat to keep the soap at a constant liquid point. Do not let the soap heat for more than 10 minutes.

2. Remove from the microwave or stovetop and stir lightly to melt completely any remaining soap pieces. Do not leave the mixing spoon in the soap while heating in the microwave or when melting on the stovetop.

3. Add the fragrance oils one drop at a time until the desired level of scent is achieved.

4. Pour the soap into the mold immediately.

5. Let the soap cool and harden completely before removing from the mold. The soap will pop out easily when completely set. For a fast set, allow the soap to set in the refrigerator until cooled.

Gold Standard

- -

MAKES 2 BARS

Gold-plated faucets might be a bit over the top for your humble digs, but there's no reason you can't bring a little luxury to your bathtub with a bar of soap. This gold-leaf bar will most certainly make you feel like a millionaire!

MATERIALS

- 3 ounces clear glycerin soap base with a pinch of gold luster powder
- 2 drops green colorant
- 5 drops pine fragrance oil
- 5 drops violet fragrance oil
- 5 drops vanilla fragrance oil
- 2 leaf motif molds, 1½ ounces each

1. Place the soap base into a heat-resistant glass measuring cup and microwave for approximately 20 to 30 seconds. If you melt the soap in a double boiler on the stove, adjust the heat to keep the soap at a constant liquid point. Do not let the soap

heat for more than 10 minutes. Meanwhile, lightly dust the embossed areas of the molds with the gold luster powder.

2. Remove from the microwave or stovetop and stir lightly to melt completely any remaining soap pieces. Do not leave the mixing spoon in the soap while heating in the microwave or when melting on the stovetop.

3. Pour 1 ounce of the soap into the embossed areas of the molds immediately.

4. Let it harden a bit, then take a piece of hard plastic (an old credit card works great) and scrape away any soap that spilled over and out of the embossed design area.

5. Add the colorant to the remaining melted soap. Stir gently to mix. If the soap starts to solidify, reheat it to remelt it.

6. Add the fragrance oils one drop at a time until the desired level of scent is achieved.

7. Pour the soap into the molds immediately.

8. Let the soap cool and harden completely before removing from the mold. The soap will pop out easily when completely set. For a fast set, allow the soap to set in the refrigerator until cooled.

Bombshell's Secret Weapon

MAKES 2 BARS

It's not easy being beautiful, as you well know. So, to ease the stress of looking good, every woman should make this soap. The blend of peppermint and orange fragrances will leave you feeling refreshed and ready to go—ready to go get a manicure!

MATERIALS
- 6 ounces clear glycerin soap base
- 3 drops blue colorant
- 3 drops green colorant
- 10 drops peppermint fragrance oil
- 15 drops sweet orange fragrance oil
- 2 clamshell molds, 3 ounces each, with embossed starfish

1. Place the soap base into a heat-resistant glass measuring cup and microwave for approximately 20 to 30 seconds. If you melt the soap in a double boiler on the stove, adjust the heat to keep the soap at a constant liquid point. Do not let the soap heat for more than 10 minutes.

2. Remove from the microwave or stovetop and stir lightly to melt completely any remaining soap

pieces. Do not leave the mixing spoon in the soap while heating in the microwave or when melting on the stovetop.

3. Add the colorant to the melted soap. Stir gently to mix. If the soap starts to solidify, reheat it to remelt it.

4. Add the fragrance oils one drop at a time until the desired level of scent is achieved.

5. Pour 1 ounce of the soap into the embossed areas of the molds immediately.

6. Let it harden a bit, then take a piece of hard plastic (an old credit card works great) and scrape away any soap that spilled over and out of the embossed design area.

7. Pour the remaining melted soap into the molds immediately.

8. Let the soap cool and harden completely before removing from the mold. The soap will pop out easily when completely set. For a fast set, allow the soap to set in the refrigerator until cooled.

Soap for Ladies Who (Wish to) Lunch

MAKES 3 BARS

While you may not be lucky enough to don a big hat and nibble watercress sandwiches, you can still treat yourself like a lady with this very proper soap. The rosemary and milk give this soap a dignified scent and creamy feel that will inspire you to bring out your pearls and pumps.

MATERIALS

- 5 ounces glycerin soap base with added olive oil and suspension formula
- ½ teaspoon dried rosemary
- ½ teaspoon powdered whole milk
- 1 teaspoon liquid glycerin
- 10 drops rosemary fragrance oil
- 3 rectangle molds, 1½ ounces each, with embossed motifs

1. Place the soap base into a heat-resistant glass measuring cup and microwave for approximately 30 to 45 seconds. If you melt the soap in a double boiler on the stove, adjust the heat to keep the soap at a constant liquid point. Do not let the soap heat for more than 10 minutes.

2. Remove from the microwave or stovetop and stir lightly to melt completely any remaining soap pieces. Do not leave the mixing spoon in the soap

while heating in the microwave or when melting on the stovetop.

3. Immediately add the dried rosemary, powdered milk, and liquid glycerin to the melted soap. Stir gently to mix. If the soap starts to solidify, reheat it to remelt it.

4. Add the fragrance oil one drop at a time until the desired level of scent is achieved.

5. Pour the soap into the molds immediately.

6. Let the soap cool and harden completely before removing from the mold. The soap will pop out easily when completely set. For a fast set, allow the soap to set in the refrigerator until cooled.

Maxed-Out Vixen Bar

MAKES 2 BARS

Not enough energy to continue being the busy lady you are? Then let the soothing blend of this soap revive you so you can continue on your merry multi-tasking way.

MATERIALS

- 1 ounce whitened glycerin soap base
- 6 ounces clear glycerin soap base
- 4 drops red colorant
- 3 drops blue colorant
- 10 drops green tea fragrance oil
- 5 drops cinnamon fragrance oil
- 2 round molds, 3 ounces each, with embossed Southwestern motifs

1. Place the whitened soap base into a heat-resist-ant glass measuring cup and microwave for approximately 10 seconds. If you melt the soap in a double boiler on the stove, adjust the heat to keep the soap at a constant liquid point. Do not let the soap heat for more than 10 minutes.

2. Remove from the microwave or stovetop and stir lightly to melt completely any remaining soap pieces. Do not leave the mixing spoon in the soap while heating in the microwave or when melting on the stovetop.

3. Immediately add the red colorant to the melted soap. Stir gently to mix. If the soap starts to solid-ify, reheat it to remelt it.

4. Pour the soap into the embossed areas of the molds immediately.

5. Let it harden a bit, then take a piece of hard plas-tic (an old credit card works great) and scrape away any soap that spilled over and out of the embossed design area.

6. Place the clear soap base into a heat-resistant glass measuring cup and microwave for approximately 10 to 25 seconds. If you melt the soap in a double boiler on the stove, adjust the heat to keep the soap at a constant liquid point. Do not let the soap heat for more than 10 minutes.

7. Remove from the microwave or stovetop and stir lightly to melt completely any remaining soap pieces. Do not leave the mixing spoon in the soap while heating in the microwave or when melting on the stovetop.

8. Immediately add the blue colorant to the melted soap. Stir gently to mix. If the soap starts to solidify, reheat it to remelt it.

9. Add the fragrance oils one drop at a time until the desired level of scent is achieved.

10. Pour the soap into the molds immediately.

11. Let the soap cool and harden completely before removing from the mold. The soap will pop out easily when completely set. For a fast set, allow the soap to set in the refrigerator until cooled.

1. Place the soap base into a heat-resistant glass measuring cup and microwave for approximately 30 to 45 seconds. If you melt the soap in a double boiler on the stove, adjust the heat to keep the soap at a constant liquid point. Do not let the soap heat for more than 10 minutes.

2. Remove from the microwave or stovetop and stir lightly to melt completely any remaining soap pieces. Do not leave the mixing spoon in the soap while heating in the microwave or when melting on the stovetop.

3. Immediately add lemongrass and the dried mint leaves to the melted soap. Stir gently to mix. If the soap starts to solidify, reheat it to remelt it.

Femme Fatale Bar

MAKES 1 BAR

The subtle blend of lemongrass and mint will give you the mystery you need to charm any man.

MATERIALS

- 2½ ounces avocado-cucumber soap base with suspension formula
- ¼ teaspoon lemongrass
- ¼ teaspoon dried mint leaves
- 2 drops lemongrass fragrance oil
- 1 hexagon massage mold, 2½-ounces

4. Add the fragrance oil one drop at a time until the desired level of scent is achieved.

5. Pour the soap into the mold immediately.

6. Let the soap cool and harden completely before removing from the mold. The soap will pop out easily when completely set. For a fast set, allow the soap to set in the refrigerator until cooled.

Potpourri Princess

While you might not be the Queen of England, you certainly deserve to treat yourself like royalty. And the best way to remind yourself of your royal roots is to experiment with the delightful potpourris in this chapter. The word potpourri comes from French, and translates literally to "rotten pot." References to potpourri recipes and functions are found in the writings of almost every culture and every time period. As cultures evolved, fragrance became a signifier of class and wealth. Once a family had created a personal potpourri recipe, everything from bed sheets to clothing to candles to jewelry, soap, and writing inks would be saturated with that fragrance. So, play around with the recommended blends or try your own, and remind the world through your personalized scent that you indeed should be treated like a princess!

Making Potpourri

All the potpourri projects may be made with either store-bought or homemade ingredients. If you enjoy gardening, you will thrill to discover that making your own potpourri is a natural way to preserve your favorite flowers and herbs. But as this first section explains, making homemade potpourri is also an option for those without a garden.

FRAGRANT MATERIALS

Fragrant plants and spices have always been and still are the most popular ingredients in potpourri. These materials can include any number of fragrant flowers, herbs, and spices, including roses, lavender, scented geranium, basil, star anise, and cinnamon, to name a few.

DECORATIVE MATERIALS

As fragrance oils have become more accessible and reasonably priced, decorative, unscented materials have become popular potpourri ingredients. Colorful whole blossoms and leaves make especially interesting additions to potpourris that are displayed in bowls or crafts.

WOOD SHAVINGS

Large shavings of wood are abundant in commercially made potpourris because they are inexpensive filler materials and absorb the scents from fragrance oils well. For homemade potpourris, the delicate curls of wood shavings from artisans are lovely additions. They blend naturally with flowers and leaves and also absorb fragrance oils well.

FIXATIVES

Since most natural fragrances have a relatively short life span, fixatives are often added to homemade potpourris. Fixatives are natural materials that absorb and hold the fragrances they are exposed to. Many fixatives have a fragrance of their own, so you will want to be sure it is compatible with the other fragrances in your potpourri. Several types of fixatives can be mixed in a potpourri as long as you find their combined fragrances pleasing. Orrisroot, gum benzoin resin, tonka beans, vetiver root, sandalwood bark, and patchouli leaves are several of the most popular fixatives. Since you will probably have to rely on mail order and Internet sources to find them, you may want to order small quantities of each type so you can discover which ones and combinations you like best. For a homemade potpourri, you will need approximately 1 tablespoon of powdered fixative for each cup of potpourri.

FRAGRANCE OILS

If you are making potpourri in the traditional way, fragrance oils are added to the potpourri before it is left to cure. Fragrance oils are also a quick way to add fragrance to a potpourri made from colorful but unscented materials, or to refresh the fragrance of an older potpourri. Or, if you prefer the natural scent of your ingredients, you can skip fragrance oils entirely.

Where to Find Potpourri Ingredients

GROW THEM

Growing your own materials is perhaps the easiest and least expensive way to have ingredients for your potpourris always at hand, and you will find your garden a constant source of inspiration for new potpourri recipes. Surprisingly, though, some gardeners find it too painful to harvest the flowers and herbs they have nourished through the season. For these gardeners there's always the option of planting an extra bed of plants just for potpourri or collecting petals and leaves from the ground after a rain or windstorm.

SCAVENGE FOR THEM

A true potpourri lover has no shame, and will scavenge for potpourri ingredients anytime and anywhere the opportunity presents itself. When you are visiting friends and walking through their gardens, look on the ground at the base of their plants for fallen petals and seed pods. The next time you are in a florist's shop, peak into the bottom of the cooler for interesting flowers and leaves—you may be surprised at how helpful flower lovers are when they know their discarded materials will be appreciated and given new life. Plant nurseries are another place for great finds—just wander down the aisles with your eyes focused on the ground for fallen spoils. Hiking and picnic excursions are an obvious time to look for potpourri materials, although many people forget to look in the fall and winter, when seed pods and other interesting materials are in abundance.

PURCHASE THEM

A wide variety of single potpourri ingredients can be ordered in small, inexpensive quantities from mail order and Internet sources. Even if your garden is bountiful and you have had great success at scavenging, it is still fun to order materials indigenous to other parts of the country and materials with fascinating names (deer's tongue, for example) that you have never seen before.

Drying

All the materials in your potpourri will need to be completely dry. Once a plant's natural moisture is gone, the flowers and leaves will feel crispy, almost like breakfast cereal. Many flowers and leaves will shrink as they dry, and you should not be too disappointed if the brilliant colors in your fresh flowers

fade as they dry. Always pick flowers and leaves for drying on a sunny day, after the moisture from dew or rain has dried.

Most materials that are dried for fragrance can simply be spread out on paper towels and turned every few days until they are dry. This method has several drawbacks, though. If you live in a very humid environment, the materials may never dry completely, and blossoms with any weight to them may dry lopsided. Alternative methods of drying are discussed below, and since none of them require great technical knowledge or expense, you should consider experimenting with them to see which achieve the best results.

SCREEN DRYING

This method is ideal if you live in a humid area or want a bloom to dry without losing its shape. Simply elevate a piece of metal screen on bricks, books, or anything else you have handy that will allow air to circulate above and below it. Single leaves and petals should be placed flat on the screen, while whole blooms and seed heads should be positioned with stems protruding through a hole in the screen and heads facing upward. Allow at least a half inch of space around each item to ensure good ventilation. Leaves should be turned every few days.

HANG DRYING

Hang drying is a quick way to dry large amounts of sturdy materials. With this method, four to six stems of an herb or flower are tied loosely together and then hung upside down in a dark, dry area for two to four weeks. When the drying process is complete, simply separate blooms and leaves from the stems. Statice (annual, German, and caspia), love-in-a-mist, lavender, and silver king artemisia all dry well with this method.

SILICA-GEL DRYING

Drying with silica gel (and other desiccants such as sand, borax, and kitty litter) can provide stunning,

colorful blooms to top off a potpourri. Many of the flowers that will dry in desiccants just will not dry using other methods without turning brown. Do not be turned off by the initial expense—used silica gel crystals can be baked in the oven and then reused.

Dry with silica gel by sprinkling about an inch of the crystals on the bottom of a cardboard box. Position the blooms heads down on top of the silica gel so that no petals overlap, and then sprinkle another inch of silica gel over the blooms. Continue layering silica gel and flowers until the box is full or you exhaust your materials. Drying times vary, but the average range is three to seven days. Pansies, violets, zinnias, and daisies dry especially well with silica gel.

As innocent as it looks, silica gel is toxic, and care should be taken not to inhale any of the fine dust which may rise as you layer it over the flowers. Wash your hands after each exposure, and avoid the temptation to work in the kitchen. (It could easily be mistaken for sugar!) Always store your boxes of drying blooms on a high shelf, well out of reach of curious children and pets.

PUT IT ALL TOGETHER

Once your materials are dry, you are ready to begin assembling a base of fragrant materials in a bowl. You can choose materials with the same type of fra-

grance or mix and match to create a completely individual fragrance. Some commonly recognized categories of fragrance include floral fragrances (roses, scented geraniums, honeysuckle), fresh fragrances (lavender, rosemary), spicy fragrances (allspice, cinnamon, star anise, nutmeg, cloves), and citrus fragrances (lemon verbena, lemon balm, lemon basil, lemon, lime, and orange peel).

When you are happy with the fragrance of your base, it is time to make your potpourri look pretty. The materials you choose should complement the surroundings where the potpourri will be displayed. This may sound complicated, but actually it is quite fun. If the potpourri is for the dining room table and your favorite tablecloth is embroidered with delicate pansies, then you may want to arrange several pansies on the top of your bowl of potpourri. If the potpourri is for a bathroom and you have just redecorated in mauve, you will enjoy finding flowers like statice and globe amaranth that are just the right color.

The great thing about potpourri is that the proportions are entirely up to you. If you enjoy cinnamon, by all means, add more cinnamon sticks than bay leaves. Or if you have more lavender than rose petals, that is perfectly fine. In other words, you do not need a fixed amount of any ingredient to create a lovely potpourri. Start small with the fragrance oils. While it does not hurt to have a strong scent, you can always add more.

SACHETS

The popularity of the sachet dates back to early Greece, when small muslin bags were placed by each guest's plate at banquets. (This tradition continued through the centuries into the late 1700s.) Also known as "sweet bags," sachets can be filled with any combination of fragrant dried materials that you find pleasing. The sachets of Henry III were filled with violets, roses, sandalwood, cloves, coriander, and lavender, while Queen Isabella of Spain preferred coriander, orris flowers, calamus, and roses.

Historically, sachets were filled with materials that had been crushed with a mortar and pestle into a fine powder. Today's sachets are usually filled with regular potpourri, thus allowing you to rejuvenate the sachet fragrance every few months by simply squeezing the sachet.

The basic sachet is very simple to make, requiring only two side seams. (The seams can also be secured with hot glue if you would rather not sew.) Almost any fabric will work, although sachets made from natural fiber fabrics (such as cotton, silk, or linen) tend to release the fragrance better.

SLEEP PILLOWS

Sleep pillows tend to be slightly larger in size than sachets, are usually sewn shut on all sides, and can be edged with ruffles, lace, or piping. Sleep pillows are a wonderful way to add delicate fragrance around your home, and you can move them around from room to room as needed. Their small size invites additions of lace and ruffles, and the top surface is an ideal place to showcase antique laces and special needlework skills.

The first sleep pillows were made by the Romans, who filled them with dried rose petals. Through the centuries that followed, pillows were also filled with hops (to induce sleep), lavender (to fragrance a sick room), rosemary (to free the sleeper from evil dreams), and cloves (to prevent snoring).

POTPOURRI TIPS

- Keep proportions in mind when choosing display bowls and baskets: potpourri with very large ingredients look out of place in small bowls, and potpourris with very small, delicate materials look lost in large, rustic baskets.

- If you would like to display your favorite potpourri in a large bowl or basket but do not have enough to fill it, line the bottom of the container with newspaper or tissue paper and then cover completely with potpourri.

- Narrow satin ribbons make lovely accents in potpourri crafts and can be embellished with tiny love knots or small blooms from a potpourri.

- When your potpourri's fragrance begins to fade, try placing it in a moist room, such as a bathroom or kitchen. Moisture releases the natural fragrance of many plants. Or, crumble the dried materials between fingers, adding fresh ingredients and a few drops of fragrance oil.

POTPOURRI TRAPS

- Potpourris displayed on a window sill or other sunny location will quickly lose their color.

- Never display potpourri near an open window or in the path of a fan.

- Make sure all your ingredients are really dry before mixing them together or mold will develop.

- The fragrance oils used in commercially made potpourris (and some homemade potpourris) often leave stains on wood, fabric, and walls. Cotton liners can be sewn in your sachets if the fabric is thin or of great value to you, and felt lining can be applied to the bottom of display bowls or to the backs of wreaths.

- Be sure to keep display potpourris and potpourri crafts out of reach of curious toddlers.

1. Dry the rose geranium leaves, globe amaranth, and rose petals using the most appropriate method, described earlier in this chapter, for your climate and the condition of your materials.

2. After rose geranium leaves, globe amaranth, and rose petals have dried completely, mix them together in a large bowl.

3. For each cup of potpourri, sprinkle about 1 tablespoon of powdered fixative over the dry materials.

4. Add the fragrance oils to the potpourri. Be sure to use only a few drops; you can add more later if the scent is too weak.

5. Mix the potpourri well with your fingers or with a plastic or wooden spoon reserved for potpourri crafts, and place the potpourri in a brown paper bag. Shake well and roll up the bag tightly to expel the air. Place in a dark location.

6. Shake the bag once a day for a week; then shake once a week for five weeks.

7. Remove the potpourri from the bag and place in a decorative container.

Potpourri for the Palace

The wonderful rose scent will take you away from the stresses of everyday life and into a mental garden of delightful relaxation. Think Jane Austen!

MATERIALS

- rose geranium leaves
- globe amaranth
- rose petals
- fixative of your choice
- 5 parts bois de rose fragrance oil
- 3 parts lemon fragrance oil
- 2 parts geranium fragrance oil

Sweet-Smelling Sachets

Nothing is more feminine than the sweet smell of lavender nestled among your clothes. Try tucking the sachet in your lingerie drawer for a quaint pleasure.

MATERIALS
- lavender
- miniature rosebuds
- orrisroot
- fixative of your choice
- 6 parts lavender fragrance oil
- 5 parts vetiver fragrance oil
- 4 parts anise fragrance oil

1. Dry the lavender, rosebuds, and orrisroot using the most appropriate method, described earlier in this chapter, for your climate and the condition of your materials.

2. After the lavender, rosebuds, and orrisroot have dried completely, mix them together in a large bowl.

3. For each cup of potpourri, sprinkle about 1 table-spoon of powdered fixative over the dry materials.

4. Add the fragrance oils to the potpourri. Be sure to use only a few drops; you can add more later if the scent is too weak.

5. Mix the potpourri well with your fingers or with a plastic or wooden spoon reserved for potpourri crafts, and place the potpourri in a brown paper bag. Shake well and roll up the bag tightly to expel the air. Place in a dark location.

6. Shake the bag once a day for a week; then shake once a week for five weeks.

7. Create a sachet. Start with a piece of fabric that is 16¼" by 11". Fold it in half, right sides together, to form an 8⅛" by 5½" rectangle.

8. Pin the side seams together. Sew with a ¼-inch seam allowance. Turn the sachet bag right sides

out and press well. Press the top of the sachet bag under ¼ inch and sew. Then press it under another inch.

9. Fill the bag three-fourths full with potpourri and tie it closed with ribbon. Tip: If you are working with a very thin or cherished piece of fabric and your potpourri contains fragrance oils, you may want to make a small pouch out of netting and then place the pouch inside the sachet bag to prevent fabric stains.

10. The finished sachet is now ready to be decorated with satin ribbon, dried flowers, dried herbs, whole spices, gathered lace, silk flowers, or any combination of the above. All of these materials can be attached to the sachet in seconds with a glue gun.

Witch Repellant

This potpourri is perfect for fall. The aroma of apples, cinnamon, and cloves screams hot apple cider.

MATERIALS

- dried apple slices
- rose hips
- star anise
- cloves
- juniper berries
- sweet-gum seed balls and bark
- pine needles
- red peppers
- cinnamon sticks
- oak moss
- assorted seed balls and pods
- fixative of your choice
- 5 parts chamomile fragrance oil
- 3 parts spruce fragrance oil
- 2 parts citronella fragrance oil

1. Dry the dried apple slices, rose hips, star anise, cloves, juniper berries, sweet-gum seed balls and bark, pine needles, red peppers, cinnamon sticks, oak moss, and assorted seed balls and pods using the most appropriate method, described earlier in this chapter, for your climate and the condition of your materials.

2. After the dried apple slices, rose hips, star anise, cloves, juniper berries, sweet-gum seed balls and bark, pine needles, red peppers, cinnamon sticks, oak moss, and assorted seed balls and pods have dried completely, mix them together in a large bowl.

3. For each cup of potpourri, sprinkle about 1 table-spoon of powdered fixative over the dry materials.

4. Add the fragrance oils to the potpourri. Be sure to use only a few drops; you can add more later if the scent is too weak.

5. Mix the potpourri well with your fingers or with a plastic or wooden spoon reserved for potpourri crafts, and place the potpourri in a brown paper bag. Shake well and roll up the bag tightly to expel the air. Place in a dark location.

6. Shake the bag once a day for a week; then shake once a week for five weeks.

7. Remove the potpourri from the bag and place in a decorative container.

Lazy Daisy Potpourri

Instantly bring color to a room with this rainbow of pinks, purples, and creams.

MATERIALS

- zinnias
- strawflowers
- annual statice
- blue salvia
- globe amaranth
- tulips
- dusty miller
- daisies
- fixative of your choice
- 3 parts ylang-ylang fragrance oil
- 2 parts bois de rose fragrance oil

1. Dry the zinnias, strawflowers, annual statice, blue salvia, globe amaranth, tulips, dusty miller, and daisies using the most appropriate method, described earlier in this chapter, for your climate and the condition of your materials.

2. After the zinnias, strawflowers, annual statice, blue salvia, globe amaranth, tulips, dusty miller, and daisies have dried completely, mix them together in a large bowl.

3. For each cup of potpourri, sprinkle about 1 tablespoon of powdered fixative over the dry materials.

4. Add the fragrance oils to the potpourri. Be sure to use only a few drops; you can add more later if the scent is too weak.

5. Mix the potpourri well with your fingers or with a plastic or wooden spoon reserved for potpourri crafts, and place the potpourri in a brown paper bag. Shake well and roll up the bag tightly to expel the air. Place in a dark location.

6. Shake the bag once a day for a week; then shake once a week for five weeks.

7. Remove the potpourri from the bag and place in a decorative container.

Think Pink Potpourri

Indulge your inner princess with this profusion of pink.

MATERIALS

- larkspur
- zinnias
- strawflowers
- celosia
- globe amaranth
- rattail statice
- cockscomb
- achillea (The Pearl)
- tulips
- fixative of your choice
- 1 part jasmine fragrance oil
- 1 part lemongrass fragrance oil

1. Dry the larkspur, zinnias, strawflowers, celosia, globe amaranth, rattail statice, cockscomb, achillea, and tulips using the most appropriate method, described earlier in this chapter, for your climate and the condition of your materials.

2. After the larkspur, zinnias, strawflowers, celosia, globe amaranth, rattail statice, cockscomb, achillea, and tulips have dried completely, mix them together in a large bowl.

3. For each cup of potpourri, sprinkle about 1 tablespoon of powdered fixative over the dry materials.

4. Add the fragrance oils to the potpourri. Be sure to use only a few drops; you can add more later if the scent is too weak.

5. Mix the potpourri well with your fingers or with a plastic or wooden spoon reserved for potpourri crafts, and place the potpourri in a brown paper bag. Shake well and roll up the bag tightly to expel the air. Place in a dark location.

6. Shake the bag once a day for a week; then shake once a week for five weeks.

7. Remove the potpourri from the bag and place in a decorative container.

Snow White's Potpourri

White flowers immediately bring understated elegance to a room, as does this whimsical potpourri, made with white, cream, and yellow flowers.

MATERIALS
- strawflowers
- globe amaranth
- pearly everlasting
- hydrangea
- cockscomb
- rosebuds
- yarrow
- bells-of-Ireland
- fixative of your choice
- 2 parts vanilla fragrance oil
- 1 part neroli fragrance oil
- 1 part sweet orange fragrance oil
- 1 part sandalwood fragrance oil

1. Dry the strawflowers, globe amaranth, pearly everlasting, hydrangea, cockscomb, rosebuds, yarrow, and bells-of-Ireland using the most appropriate method, described earlier in this chapter, for your climate and the condition of your materials.

2. After the strawflowers, globe amaranth, pearly everlasting, hydrangea, cockscomb, rosebuds, yarrow, and bells-of-Ireland have dried completely, mix them together in a large bowl.

3. For each cup of potpourri, sprinkle about 1 tablespoon of powdered fixative over the dry materials.

4. Add the fragrance oils to the potpourri. Be sure to use only a few drops; you can add more later if the scent is too weak.

5. Mix the potpourri well with your fingers or with a plastic or wooden spoon reserved for potpourri crafts, and place the potpourri in a brown paper bag. Shake well and roll up the bag tightly to expel the air. Place in a dark location.

6. Shake the bag once a day for a week; then shake once a week for five weeks.

7. Remove the potpourri from the bag and place in a decorative container.

Sweet Dreams Sleep Pillow

Rosemary is said to help memory, so hopefully this blend of lovely aromas will help you remember the good things in life.

MATERIALS

- peppermint
- spearmint
- rosemary
- lemon balm
- honesty
- Christmas roses
- fixative of your choice
- 5 parts spearmint fragrance oil
- 3 parts copaiba fragrance oil
- 2 parts fennel fragrance oil

1. Dry the peppermint, spearmint, rosemary, lemon balm, honesty, and Christmas roses using the most appropriate method, described earlier in this chapter, for your climate and the condition of your materials.

2. After the peppermint, spearmint, rosemary, lemon balm, honesty, and Christmas roses have dried completely, mix them together in a large bowl.

3. For each cup of potpourri, sprinkle about 1 tablespoon of powdered fixative over the dry materials.

4. Add the fragrance oils to the potpourri. Be sure to use only a few drops; you can add more later if the scent is too weak.

5. Mix the potpourri well with your fingers or with a plastic or wooden spoon reserved for potpourri crafts, and place the potpourri in a brown paper bag. Shake well and roll up the bag tightly to expel the air. Place in a dark location.

6. Shake the bag once a day for a week; then shake once a week for five weeks.

7. Create a sleep pillow. Start with two fabric rectangles, both 10¾" by 8½". Or you can make your own pattern by tracing the shape of a box or a plate, if you want a round pillow.

8. Cut 2 to 4 layers of quilt batting the same size as the pillow pieces and pin them to the wrong side of the pillow's bottom piece. Experiment with the number of batting layers until you are happy with the thickness.

9. Pin the two pillow pieces together, right sides together, and sew the seams with a ¼-inch seam allowance. Leave a 2- to 3-inch opening. Turn the pillow right sides out and press.

10. Cut two layers of tulle or netting slightly smaller than the pillow. Top-stitch the tulle together on three sides, fill with potpourri, and then top-stitch the remaining side.

11. Slide this potpourri pouch into the pillow and sew the opening closed by hand. This seam can be reopened later to launder the pillow or to add a fresh pouch of potpourri.

12. To add a ruffle, cut a length of lace or fabric twice as long as the pillow's sides. Gather the lace or fabric with a running stitch (or use pregathered lace). Hem the ruffle if needed. Before sewing the pillow, position the lace between the two pieces of fabric with the gathered edge flush with the outer edges of the fabric. Sew seams as usual.

A Pillow Fit for Sleeping Beauty

This fragrant aroma will lull you to sleep faster than you can count sheep.

MATERIALS
- bee balm
- rose petals
- lavender
- peppermint
- comfrey
- lemon verbena
- chamomile
- fixative of your choice
- 2 parts sweet orange fragrance oil
- 1 part vanilla fragrance oil
- 1 part petitgrain fragrance oil

1. Dry the bee balm, rose petals, lavender, peppermint, comfrey, lemon verbena, and chamomile using the most appropriate method, described earlier in this chapter, for your climate and the condition of your materials.

2. After the bee balm, rose petals, lavender, peppermint, comfrey, lemon verbena, and chamomile have dried completely, mix them together in a large bowl.

3. For each cup of potpourri, sprinkle about 1 tablespoon of powdered fixative over the dry materials.

4. Add the fragrance oils to the potpourri. Be sure to use only a few drops; you can add more later if the scent is too weak.

5. Mix the potpourri well with your fingers or with a plastic or wooden spoon reserved for potpourri crafts, and place the potpourri in a brown paper bag. Shake well and roll up the bag tightly to expel the air. Place in a dark location.

6. Shake the bag once a day for a week; then shake once a week for five weeks.

7. Create a sleep pillow. Start with two fabric rectangles, both 10¾" by 8½". Or you can make your own pattern by tracing the shape of a box or a plate, if you want a round pillow.

8. Cut 2 to 4 layers of quilt batting the same size as the pillow pieces and pin them to the wrong side of the pillow's bottom piece. Experiment with the number of batting layers until you are happy with the thickness.

9. Pin the two pillow pieces together, right sides together, and sew the seams with a ¼-inch seam allowance. Leave a 2- to 3-inch opening. Turn the pillow right sides out and press.

10. Fill the pillow with small pieces of fiberfill and potpourri until you are happy with the fullness. Sew the opening closed by hand.

11. To add a ruffle, cut a length of lace or fabric twice as long as the pillow's sides. Gather the lace or fabric with a running stitch (or use pregathered lace). Hem the ruffle if needed. Before sewing the pillow, position the lace between the two pieces of fabric with the gathered edge flush with the outer edges of the fabric. Sew seams as usual.

1. Dry the bee balm blooms, scented geranium leaves, and peppermint using the most appropriate method, described earlier in this chapter, for your climate and the condition of your materials.

2. After the bee balm blooms, scented geranium leaves, and peppermint have dried completely, mix them together in a large bowl.

3. For each cup of potpourri, sprinkle about 1 tablespoon of powdered fixative over the dry materials.

4. Add the fragrance oils to the potpourri. Be sure to use only a few drops; you can add more later if the scent is too weak.

5. Mix the potpourri well with your fingers or with a plastic or wooden spoon reserved for potpourri crafts, and place the potpourri in a brown paper bag. Shake well and roll up the bag tightly to expel the air. Place in a dark location.

6. Shake the bag once a day for a week; then shake once a week for five weeks.

7. Create a sachet. Start with a piece of fabric that is 16¼" by 11". Fold it in half, right sides together, to form an 8⅛" by 5½" rectangle.

Prince-Charming-Come-Hither Sachet

Hang this sachet in your closet, and you will always have a contented spirit as your clothing picks up this peaceful scent.

MATERIALS
- bee balm blooms
- scented geranium leaves
- peppermint
- fixative of your choice
- 5 parts peppermint fragrance oil
- 3 parts sandalwood fragrance oil
- 2 parts sweet basil fragrance oil

8. Pin the side seams together. Sew with a ¼-inch seam allowance. Turn the sachet bag right sides out and press well. Press the top of the sachet bag under ¼-inch and sew. Then press it under another inch.

9. Fill the bag ¾ full with potpourri and tie it closed with ribbon. Tip: Don't use fine fabric. See previous note.

10. The finished sachet is now ready to be decorated with satin ribbon, dried flowers, dried herbs, whole spices, gathered lace, silk flowers, or any combination of the above. All of these materials can be attached to the sachet in seconds with a glue gun.

Enchanting Potpourri

The lovely combination of jasmine, vanilla, and lavender will ease your troubles and create a blissful mood.

MATERIALS
- jasmine
- globe amaranth foliage (flowers, leaves, and stems)
- mountain mint seed heads
- black-eyed Susan seed heads
- fixative of your choice
- 2 parts jasmine fragrance oil
- 1 part vanilla fragrance oil
- 1 part lavender fragrance oil

1. Dry the jasmine, globe amaranth foliage, mountain mint seed heads, and black-eyed Susan seed heads using the most appropriate method, described earlier in this chapter, for your climate and the condition of your materials.

2. After the jasmine, globe amaranth foliage, mountain mint seed heads, and black-eyed Susan seed heads have dried completely, mix them together in a large bowl.

3. For each cup of potpourri, sprinkle about 1 tablespoon of powdered fixative over the dry materials.

4. Add the fragrance oils to the potpourri. Be sure to use only a few drops; you can add more later if the scent is too weak.

5. Mix the potpourri well with your fingers or with a plastic or wooden spoon reserved for potpourri crafts, and place the potpourri in a brown paper bag. Shake well and roll up the bag tightly to expel the air. Place in a dark location.

6. Shake the bag once a day for a week; then shake once a week for five weeks.

7. Remove the potpourri from the bag and place in a decorative container.

Serene Scent

Renew yourself with this pleasing blend of scents and artful smattering of colors.

MATERIALS

- strawflowers
- roses
- bachelor's buttons
- larkspur
- lemongrass
- pepperberries
- orrisroot
- sunflower petals, head, and seeds
- lemon verbena leaves
- fixative of your choice
- 1 part lemon fragrance oil
- 1 part lavender fragrance oil

1. Dry the strawflowers, roses, bachelor's buttons, larkspur, lemongrass, pepperberries, orrisroot, sunflower petals, head, and seeds, and lemon verbena leaves using the most appropriate method, described earlier in this chapter, for your climate and the condition of your materials.

2. After the strawflowers, roses, bachelor's buttons, larkspur, lemongrass, pepperberries, orrisroot, sunflower petals, head, and seeds, and lemon verbena leaves have dried completely, mix them together in a large bowl.

3. For each cup of potpourri, sprinkle about 1 tablespoon of powdered fixative over the dry materials.

4. Add the fragrance oils to the potpourri. Be sure to use only a few drops; you can add more later if the scent is too weak.

5. Mix the potpourri well with your fingers or with a plastic or wooden spoon reserved for potpourri crafts, and place the potpourri in a brown paper bag. Shake well and roll up the bag tightly to expel the air. Place in a dark location.

6. Shake the bag once a day for a week; then shake once a week for five weeks.

7. Remove the potpourri from the bag and place in a decorative container.

Captivating Potpourri

This festive potpourri is fabulous for winter. It offers a pleasing scent and much needed color against dreary winter skies.

MATERIALS

- canella berries
- juniper
- lamb's ear
- goldenrod
- allium
- fixative of your choice
- 3 parts sweet orange fragrance oil
- 2 parts rosewood fragrance oil
- 2 parts frankincense fragrance oil

1. Dry the canella berries, juniper, lamb's ear, goldenrod, and allium using the most appropriate method, described earlier in this chapter, for your climate and the condition of your materials.

2. After the canella berries, juniper, lamb's ear, goldenrod, and allium have dried completely, mix them together in a large bowl.

3. For each cup of potpourri, sprinkle about 1 tablespoon of powdered fixative over the dry materials.

4. Add the fragrance oils to the potpourri. Be sure to use only a few drops; you can add more later if the scent is too weak.

5. Mix the potpourri well with your fingers or with a plastic or wooden spoon reserved for potpourri crafts, and place the potpourri in a brown paper bag. Shake well and roll up the bag tightly to expel the air. Place in a dark location.

6. Shake the bag once a day for a week; then shake once a week for five weeks.

7. Remove the potpourri from the bag and place in a decorative container.

Cheating Heart Blend

The flowers in this potpourri are reminders of infidelity, such as fickleness, distrust, grief, despair, and insincerity. Whip up a batch of this potpourri to remind your lover of his cheating heart!

MATERIALS

- pink larkspur
- purple larkspur
- lavender
- love-in-a-mist
- marigolds
- maple leaves
- foxglove
- fixative of your choice
- 3 parts patchouli fragrance oil
- 1 part rose fragrance oil
- 1 part sandalwood fragrance oil

1. Dry the pink and purple larkspur, lavender, love-in-a-mist, marigolds, maple leaves, and foxglove using the most appropriate method, described earlier in this chapter, for your climate and the condition of your materials.

2. After the pink and purple larkspur, lavender, love-in-a-mist, marigolds, maple leaves, and foxglove have dried completely, mix them together in a large bowl.

3. For each cup of potpourri, sprinkle about 1 tablespoon of powdered fixative over the dry materials.

4. Add the fragrance oils to the potpourri. Be sure to use only a few drops; you can add more later if the scent is too weak.

5. Mix the potpourri well with your fingers or with a plastic or wooden spoon reserved for potpourri crafts, and place the potpourri in a brown paper bag. Shake well and roll up the bag tightly to expel the air. Place in a dark location.

6. Shake the bag once a day for a week; then shake once a week for five weeks.

7. Remove the potpourri from the bag and place in a decorative container.